REGIONAL PLANNING AND DEVELOPMENT IN EUROPE

Regional Planning and Development in Europe

Edited by
DAVID SHAW
University of Liverpool, UK

PETER ROBERTS
University of Dundee, UK

JAMES WALSH
National University of Ireland, Maynooth, Ireland

Routledge
Taylor & Francis Group

LONDON AND NEW YORK

First published 2000 by Ashgate Publishing

Reissued 2018 by Routledge
2 Park Square, Milton Park, Abingdon, Oxon OX14 4RN
711 Third Avenue, New York, NY 10017, USA

Routledge is an imprint of the Taylor & Francis Group, an informa business

A Library of Congress record exists under LC control number: 00132806

ISBN 13: 978-1-138-74051-8 (hbk)
ISBN 13: 978-1-138-74047-1 (pbk)
ISBN 13: 978-1-315-18344-2 (ebk)

Contents

List of Figures

List of Tables

List of Contributors

Mark Baker currently works for the Government Office for the North West. Prior to this appointment he was lecturer in the School of Architecture, Planning and Landscape at the University of Newcastle upon Tyne. He is a chartered town planner with previous work experience in local and central government and is currently undertaking research into various aspects of regional and strategic planning in the UK, including a DETR sponsored study of the English structure planning process.

Wilhelm Benfer heads the Office of Economic Development and Planning of Barnim County in Germany's State of Brandenburg. Before entering the field of practice he already spent his academic studies mainly on issues of local and regional economic development and of spatial planning as they pertain to Germany, the United Kingdom and the United States.

Gordon Dabinett is a Reader in the School of Urban & Regional Studies/Centre for Regional Economic and Social Research (CRESR) at Sheffield Hallam University. He has been involved in a number of evaluation studies of UK national regeneration policies and EU Regional Programmes.

Dimitri Economou is an Associate Professor of Planning at the University of Thessaly and a former researcher in the National Centre of Social Science Research of Greece. He has worked as a consultant and written widely on urban and regional analysis and planning issues. His main current research interest is spatial policy, in a comparative European/international perspective.

David Gibbs is Professor of Human Geography in the Department of Geography at the University of Hull. His main research interests are in local and regional economic development, particularly in relation to environmental and technological policies. He is currently researching into governance and regulation issues in local environmental policy making with Andrew Jonas.

Tony Jackson is a Senior Lecturer in Environmental Management at the School of Town and Regional Planning, University of Dundee. His research interests include regional and local labour markets, environmental economics and taxation and disaster planning

Andrew Jonas is a Senior Lecturer in Geography at Hull University. He has published widely on urban politics and governance in the U.S.A.. He is currently researching conservation policy and planning in Southern California with researchers at the University of California, and local environmental policy making in the U.K.. He is co-editor of "The Urban Growth Machine: Critical Perspectives Two Decades Later" (State University of New York Press).

Vince Nadin is a Reader and the Director of the Centre for Environment and Planning at the University of the West of England. Vince's research interests lie in the field of statutory land use planning, European comparative planning and the development of European spatial policy and its impact on planning systems and planning practice.

Peter Ramsden is the Head of Brussels office for Enterprise plc where his team carries out a wide range of economic development consultancies on local and community economic development, innovation and information society issues. His group have been developing policy on fund based approaches to venture capital, property and micro credit. Before working for Enterprise, Peter was a detached national expert with DG16 of the European Commission where he managed the Merseyside Objective 1, Greater Manchester Lancashire and Cheshire Objective 2 and Yorkshire and Humberside Objective 2 programmes.

Tim Richardson is a Research Fellow in the Arkleton Centre for Rural Development Research at the University of Aberdeen. He previously undertook research at the University of Manchester which examined the 'new realism' within UK transport planning, and he is also undertaking PhD research at Sheffield Hallam University into the EU Trans-European Networks policy process.

Peter Roberts is Professor of European Strategic Planning at the Geddes Centre for Planning Research, University of Dundee. He is also Chair of the Town and Country Planning Association. Peter's research interests include regional planning and development, sustainable development concerns and strategic territorial planning, and the evolution and performance of European Union regional, spatial and environmental policy.

Heinrich Seul is co-founder and president of CREAM Consultants. He has developed cross-sectoral solutions for environment, agriculture and food markets to numerous projects of major development banks, international companies, institutions and NGOs in Europe, the Americas and Africa.

Dave Shaw is a Senior Lecturer in the Department of Civic Design, University of Liverpool. His research interests include environmental planning and management, rural planning, comparative planning and the evolution of European Union spatial development policies and their impacts on planning systems and practice.

Marinella Terrasi teaches Regional Economics in the Department of Economics at the University of Pisa, Italy. She previously worked at the Centre for the Development of Southern Italy, Naples and at the University of Sienna. She has researched and written widely on regional development and the location process of manufacturing industries during different phases of Italian post-war development.

Ivan Turok is Professor of Urban Economic Development at the University of Glasgow. His research interests include urban and regional development processes and policies, particularly in Europe. He is currently leading an ESRC-funded integrated case study of economic competitiveness and social cohesion in Scottish cities. Ivan has recently published 'The Jobs Gap: Labour Market Trends in British Cities' (The Policy Press, 1999) and 'The Coherence of EU Regional Policy: Contrasting Perspectives on the Structural Funds' (edited with John Bachtler) (Jessica Kingsley, 1997).

Christiane Wellensiek is co-founder of **CREAM** Consultants. She develops integrated solutions for sustainable development in rural areas, with special emphasis on the situation of rural women. As specialist for environmental education she has worked for many government and non-governmental organizations in Germany and France.

James Walsh is Professor and Head of the Department of Geography, NUI Maynooth. He has written extensively on regional and local development issues in Ireland. He has made several contributions to the "Ireland National Development Plan 2000-2006". Between 1994 and 1998 he was Chairman of the Irish Branch of the Regional Studies Association. Recent publications include co-edited volumes on "Innovation, Competitiveness and Regional Development in Ireland" (1997) and "Sustainable Development on the North Atlantic Margin" (Ashgate, 1997) and he is co-author of "Irish Agriculture in Transition - a census Atlas" (1999).

Preface

This book represents the output of a meeting of the European Urban and Regional Research Network held at Frankfurt-am-Oder in 1997. Many of the papers presented at the meeting have been reworked and extended in order to incorporate a wider span of material than that presented at the Conference. As editors we are grateful to the authors for their assistance in the various processes of preparation and for their patience during the past months while the text was being finalised.

We owe a considerable debt of thanks to a number of individuals and organisations. The Regional Studies Association - the organisers of the Frankfurt Conference - have given permission for this selection of papers to be published and Sally Hardy, the Director of the Association, has helped ease the way. Our colleagues at Dundee, Liverpool and Maynooth have provided considerable support and technical assistance. Special thanks are due to Jill Gun-Why, Claire LaRoc and Pam Gibb for their help in preparing individual chapters and the layout of the manuscript.

Finally we are also wish to thank our publisher Ashgate. As in many such projects, the timetable for the preparation of this book has stretched, and Anne Keirby, Sarah Markham and Val. Rose have gladly changed our publishing requirements. Their help is much appreciated and has been an essential element in ensuring the completion of this project.

Peter Roberts, David Shaw and Jim Walsh
January 2000

PART 1:

INTRODUCTION: THEMES AND CONTENT

1 Regional Planning and Development in Europe: An Introduction

PETER ROBERTS, DAVID SHAW AND JAMES WALSH

The Context

Although many assessments of the recent evolution of regional planning and development policy in Europe have placed emphasis on the growing coincidence of theory and practice across countries and sectors of the economy, many individual and distinctive features remain. These reflect the variations that exist in the structure of government, the success or failure of particular initiatives and the inherited characteristics of regions. These variations are important, because they represent a continuing need to balance the desirability of promoting a common European agenda and approach, against the specific characteristics, problems, needs and opportunities evident in individual regions. Whilst subsidiarity as a principle has now been accepted and introduced as a fundamental requirement into the basic European Union legislation, considerable differences persist in the practice of devolution at national level and in the operation of systems of regional management and governance.

Furthermore, a number of new elements and aspects of policy have been introduced in recent years that are only now beginning to impact on regional planning and development. Included in the list of these new policies is the increased emphasis that has been placed on sustainable development, the strengthening of competition and single market policy and its enforcement, the 'reform' of the overall purpose and operational procedures associated with the Structural Funds and Common Agriculture Policy, the introduction of European spatial policy, and the establishment of a European single currency. Although the impacts of some of these policies will take time to become fully evident at a regional level, in other cases the relationship between the introduction of a new policy and the consequences at regional level is more direct and immediate.

By suggesting that twin forces are at work - towards greater conformity, on the one hand, and in support of greater diversity, on the other hand – the intention is to underline the importance that should be attached to identifying and understanding the origins and outcomes of regional planning and development policy. As noted above, the social, political and economic inheritance of a nation or region would appear to be significant factors in determining both the present situation and future prospects.

However, and as will be illustrated later in this book, it is possible to compensate for previous regional underperformance and to even out economic and social variations – indeed, this assumption lies at the heart of regional policy. It is possible to bring about a transformation in the fundamental political and administrative capability of a region. In part, these change characteristics reflect the new political geography, or geographies of post-Soviet Union Europe, but they also represent the outcome of conscious actions that have been implemented in order to provide greater accessibility, enhanced mobility and a clearer sense of belonging to a 'single' Europe.

Key Themes and Issues

In addition to the general themes which were discussed in the previous section, a number of other more specific issues can be identified that represent the core concerns that are investigated and discussed in this book. Each of the cross-cutting themes is introduced here, but is developed in greater detail elsewhere; the themes include:

- the desirability of establishing policy regimes that reflect and meet the requirements of individual regions;
- in the absence of the above, the adaptation of an existing policy regime in order to meet the needs of a region;
- the general search for greater efficiency and 'value for money' in the design and implementation of regional policy;
- the search for mechanisms and procedures that allow for the closer integration of the various strands and elements of spatial policy and which reflect the requirements associated with the preparation of a European spatial framework;
- the importance of tracing and assessing the impacts and consequences of previous rounds of regional policy, and of evaluating the possible

implications of proposed changes in policy and methods of implementation;

- the identification and dissemination of good and best practice in regional planning and development;
- an attempt to compare and contrast the differing structures and political systems evident in the various nation states, and to identify the implications of such structures for central-regional-local relations;
- the implications of introducing new tasks and requirements at regional level, and the evaluation of the procedures adopted to discharge such tasks.

Policy Regimes

The first two themes are explored here. First, the operation of policy regimes that are tailor made to fit the requirements of an individual region, and, second, the approach used to adapt a policy in order to ensure that it can be altered in order to fit the requirements of a specific region. These are important issues, for it is now recognised that the absence of an adequate 'institutional infrastructure', which can be used to help to design and implement appropriate policy, represents a major obstacle to the successful development of a region. In the view of Wiehler and Stumm (1995), because of growing locational competition between nations and (especially) regions, territorial authorities at regional level "need sufficient powers of their own to shape their economic future" (P.249).

Some of the questions raised by a consideration of the two themes are addressed in this book. At a national-regional level, Mark Baker examines the operation in six English regions of a what is in theory a uniform national development plan system. Hardly surprisingly, the results of the inter-regional comparisons demonstrate the important role that is performed by local and regional considerations in modifying national priorities and shaping them in order better to reflect local circumstances. This is seen by Baker to be both legitimate and desirable, given that local and regional authorities have their own roles and aspirations as well as acting as the implementing agents of central government. Furthermore, this sort of approach can be supported on the grounds that local democratic control can and should be exercised over matters that are held to account at the level of the locality or community. What is more worrying, is that the reality of the national-regional-local relationship in the UK for the case of development plans, is one in which the centre has traditionally exercised a high degree of control, and, at least at the time that Baker's research was conducted, continued to exercise such control.

As Baker argues – and this coincides with the prescription offered by Wiehler and Stumm – the real solution is to be found through the devolution of central government powers and responsibilities. This, he suggests, would allow for more effective and relevant development plans to be prepared and for democratic control to be exercised over the form and content of such plans.

In another contribution, Turok examines the role of local development policies as a way of developing plans that are appropriate to the requirements of specific groups. This contribution demonstrates a number of ways of bringing together a partnership of relevant people and organisations, of developing a locally-rooted partnership programme that adapts and draws down on national policy regimes, and which allows for the introduction of 'bottom up' policies that reflect local needs. Turok argues that such an approach is now beginning to become accepted by central government, and that this belated recognition offers real potential for future policy improvement.

These two examples demonstrate both the significance of measuring progress against a standard national policy regime, and of ensuring that, in providing such an assessment, the local and regional dimension is taken into account. What is most interesting, is that elsewhere in the book other authors reflect on the same theme and come to a similar conclusion. Quite clearly, the desirability of providing either a 'tailor made' or a heavily 'tailored to fit' regime can be supported by the available research evidence.

Value for Money

Most governmental organisations, be they supranational, national, regional or local, have increasingly demonstrated a concern with ensuring that the operation of the public policy system delivers an effective standard of service in the most efficient manner. This objective is frequently translated into the search for a solution that offers good value for money. However, although value for money is the aspiration, it is clear that only a few organisations have a precise idea of what this really means and how previous eras of policy have performed.

As a consequence, many new policies or new elements of policy are introduced without the benefit of the results of either an ex-post assessment of the effects and value for money of previous rounds of policy, or an ex-ante evaluation of the likely future implications of a new policy. Two of the authors who have contributed to this book attempt to rectify these failings by looking at the past performance of regional policy and its effects (Walsh) and by evaluating the potential benefits of introducing new policy instruments (Ramsden). Although very different in approach and content, these two chapters represent two sides of the same coin. On the one hand, an assessment is presented of what has worked, why and how, and, on the other hand, an analysis is advanced of what can be done to squeeze more value out of a static or shrinking pot of public resources.

Walsh examines the case of Ireland – the 'Celtic Tiger' – and tracks the development of regional policy from the 1980s to the present day. This chapter illustrates the merits of establishing clear and specific strategic regional policy and continuing to operate such a policy over the long-hand. Despite a series of new challenges and the occasional difficulty, the overall performance of Irish regional development has been impressive. Most importantly, the Irish case demonstrates that weak and underdeveloped regions can undergo transformation, and that a blend of policies is required in order to bring about the desired results.

Looking forward rather than back, Ramsden assesses the characteristics and advantages of non-grant financial instruments. The intention of such funds is to extend the range of instruments available for regional development and to attract a higher input of resources from non-traditional public sector and private sector sources. In particular, the aim of these new instruments is to increase the availability of risk capital and venture funds, thereby allowing the SME sector to gain access to funds that would otherwise be unavailable to them. Although such funds are not without risk, and although it takes time for recipients to become familiar with and accept such novel ways of supporting investment, the new funds offer enhanced value for money in terms of public sector investment in support of regional development.

Integration of Policies

In attempting to secure enhanced value for money, many governments have turned to policy integration as a means of achieving more with the same level of resource input. Whilst this is a valid strategy from a financial perspective, it only provides part of the explanation behind the search for policy integration. Other elements of the integration process relate to the desirability of establishing a corporate framework that can be used to guide all investment in a region and the importance, especially in relation to those aspects of activity that cannot easily be physically restricted to a single region, such as atmospheric pollution, and of establishing harmony between the various fields of policy, including economic development, land use planning and environmental management.

Many of the chapters in this book consider one or more aspects of the processes and outcomes that are associated with attempts better to integrate disparate areas of activity. However, a number of the chapters pay particular attention to either the integration of two or more outputs of policy, or to the development and use of a spatial framework as an organising and integrating concept. In particular, with regard to the latter item, two chapters examine spatial planning, one examines developments at a European scale (Nadin and Shaw) whilst the other is chiefly concerned with questions related to strategic processes and procedures (Dabinett and Richardson).

The European Spatial Development Perspective (ESDP) is now a well-established element of European Union policy, and the intention in preparing the ESDP was to provide a greater sense of spatial co-ordination at all levels of activity (Ministers Responsible for Spatial Planning, 1998). In one sense, this framework setting is itself instructive as to how spatial integration can support greater economic efficiency, whilst, in another sense, the establishment of a spatial framework can be seen as an end in itself. It provides a 'gameboard' upon which the disposition of other policies and instruments can be contemplated. Nadin and Shaw outline the origins and evolution of EU spatial development policy and illustrate this history and explanation by reference to the uses to which it has and can be put. As well as considering spatial policy at the European level, the authors also consider issues related to national and regional spatial policies and the assignment of competence within the context of the debate on subsidiarity.

Dabinett and Richardson extend the discussion of policy integration through spatial planning by exploring the alternative planning paradigms which underpin the various strategic planning and policy systems. In particular, they point to the contested nature of the spatial playing field of action, the structure and consequences of the various approaches used to promote change, and the role of evaluation as a force for mediation between the competing approaches. As the authors note in their conclusion, there are interesting tensions evident between the various systems and approaches and this makes for difficulty in attempting to apply a single model of evaluation.

Two other chapters explore specific aspects of policy integration, between two or more elements of policy and in relation to the needs of particular structures for regional planning and development. Jackson and Roberts examine the case of environment and economic policy integration through the medium of Structural Fund regional programmes. This evaluation makes specific reference to the ways in which environmental 'drivers' can be used to enhance the content and performance of economic development programmes, and to the desirability of promoting the acceleration of environment – economy integration through the adoption of the 'ecological modernisation' paradigm.

In the other chapter related to this theme, Gibbs and Jonas explore the development and operation of local environmental policy making and the various ways which exist for incorporating spatial concerns into environmental policy. The conclusion is of particular importance. In the long-term it is possible for local economies to attain some level of self-regulation. Furthermore, it provides a basis for the authors to suggest that there is a case to support the adoption of a multi-layered approach to the construction of policy agendas.

Assessing Policy Performance

As well as considering the operation of policy frameworks which seek to co-ordinate or integrate two or more policy fields, there is also a very real need for research that examines the performance of policy systems as a whole. Two examples of such an approach are discussed here; the first (by Walsh) has already been referred to above, whilst the second (by Roberts) offers an assessment of the different ways in which Structural Fund regional programmes are structured and organised in different regions.

As noted already, the central contribution made by Walsh is to provide a longitudinal analysis of the performance of regional economic policy in Ireland. The key messages have already been discussed, but there is also an important contribution in the chapter than can be mobilised in order to inform the design of appropriate policies in the future. As Walsh notes, one of the fundamentals of designing an effective set of regional development policies is the establishment of a model of governance that allows for collaboration and partnership between the state, the private sector and local communities. In addition to these constitutional lessons, the chapter also points to the need to adjust policy over time, especially in order to ensure that the package as a whole conforms to the requirements of sustainable development.

Roberts examines a number of factors and key issues which have helped to mould the very different stances adopted to a common policy regime (the Structural Fund regional programme regulations) by the various member states and, in some cases, by the different regions within an individual member state. Many of these differences can be explained by reference to the varied constitutional and administrative inheritances evident in member states. Reflecting Wiehler and Stumm's (1995) analysis, the range of circumstances – from advanced federal states to countries in which centralisation dominates – is reflected in the method adopted for the incorporation of a Structural Fund programme in a suite of regional policies. Weak regionalism and the absence of regional governance is associated with the presence of multiple administrative regimes and, in some cases, the imposition of a standard national approach that limits the 'absorption' of a Structural Fund programme into the prevailing culture and practice of regional planning and development.

Good Practice

One of the reasons for conducting policy assessment is to identify good or best practice in the hope that the lessons will be disseminated within a region or to other regions. Many aspects of good practice can be seen in the chapters presented herein. For example Baker points to the most effective ways for central-local development plan preparation to proceed, whilst Walsh clearly demonstrates the fine points of regional policy design and implementation.

Above and beyond these instances, two chapters offer a number of specific examples of the construction and operation of good practice in regional planning and development. Economou considers the regional impact of Community Agricultural Policy (CAP) and demonstrates by reference to the case of Greece, how CAP resources can best be used in support of regional restructuring. Although the eventual impact of CAP transfers is low, the procedures used to co-ordinate CAP expenditure and regional policy represent good policy.

Seul and Wellensiek present a second chapter that explores a number of aspects of good practice. This examines rural development and, in particular, the ways in which sustainable development policy is constructed and implemented in such areas. Specific points of good practice that are worthy of replication include the need for both 'bottom-up' and 'top-down' action, the importance of selecting and applying appropriate targets and indicators of success, and the desirability of linking the environmental dimension of sustainability to socio-economic progress.

Different Political and Administrative Systems

The wide range of different national-regional arrangements present in Europe has resulted in the adoption of a wide variety of different approaches to the construction and implementation of policies for regional planning, development and management. The real question here is what are the characteristics of the various systems and what benefits or disbenefits are associated with the operation of the system? Many clues exist in the chapters in this book that can help to provide an answer to this question. Three of the chapters place particular emphasis on this theme.

Terrasi compares and contrasts the operation of policies designed with the intention of promoting regional convergence in Italy and Spain. Her analysis demonstrates the interactions between a number of policy fields and factors – economic development, locational disadvantage, the structure and powers of regional government, and the extent to which firms participate in international markets – and the outcomes of these processes. The conclusion is striking and important – that a country may be successful in the aggregate, without solving its regional problems!

Roberts, who compares the operation of the Structural Fund programmes and partnerships across five European Union member states, further extend this analysis.

New Tasks and Challenges

The European regional planning and policy agenda would appear to be in a constant state of flux; the drivers of change originate at various levels in the spatial hierarchy – European, national and regional – and reflect the relative priority given to a number of different areas of policy. Whilst many of the chapters examine certain aspects of these constant processes of adjustment in regional policy, two chapters focus attention on specific elements in the transformation of regional planning and policy.

Benfer offers an insight into the emerging process of co-regional planning. Here the intention is to encourage the exchange of ideas and experience between regions in different countries. The aim is to help in the identification of the effects of various levels of policy and to exchange theories, strategies and good practice. In conclusion, Benfer argues the case for the extension and adoption of the co-regional planning model on the grounds that it offers a means of both influencing European Union policy and of operationalising individual regional strategies.

Nadin and Shaw present the second case. They review and assess the growing interest in transational co-operation in spatial planning and explore the mechanisms through which such co-operation can be secured. The importance of their analysis can be seen in the evidence which they cite in support of transnational planning and, perhaps more importantly, in the case that they advance for the introduction of the subsidiarity principle into the discipline of spatial planning. This principle, which is much abused and little understood, offers guidance in the selection of appropriate fields of action at various spatial levels.

The Structure of This Book

The final task of this introductory chapter is to outline the approach adopted in the remainder of this book.

The three parts of the book reflect a number of areas of interest and types of assessment. Part I focuses on the operation of regional and associated policies, and provides evidence of the success of failure of individual national, regional and local programmes of implementation. The second part of the book includes a number of chapters organised around the theme of spatial planning, environmental management and regional development policy. In Part III, the various contributions offer a number of assessments and evaluations of past experience and provide a glimpse of the future.

The contributions contained in the following pages can only hope, at best, to offer a snapshot in time that reflects a selection of the more important aspects of regional planning and development in Europe. Other projects and publications offer other views and experiences. However, when taken together, these various contributions may help to inform future rounds of policy and practice, and will therefore have served their purpose.

References

Ministers Responsible for Spatial Planning (1997), *European Spatial Development Perspective*, Ministers Responsible for Spatial Planning, Noordwijk.
Wechler, F. and Stumm, T. (1995), 'The Powers of Local and Regional Authorities and Their Role in the European Union', *European Planning Studies*, 3, pp.227-250.

PART II:

THE OPERATIONS OF THE STRUCTURAL FUNDS AND REGIONAL RESTRUCTURING: APPROACHES AND EXPERIENCES

2 The Regional Impact of the Community's Agricultural Policy: The Case of Greece

DIMITRI ECONOMOU

Introduction

It is well known that Community Agricultural Policy (CAP) is the most important EU policy in financial terms. For many years the cost of CAP accounted for about 60% of the Community budget expenditure (EC, 1994a) and, despite the relative decline during the 1990s, it preserves its predominance in the budget, with a share of around 45%. In comparison, the budget allocation for regional policy used to be significantly lower, less than 1/3 of CAP, and only lately has it caught up with that of CAP.

The first priority of CAP is the support of agricultural income, an objective which, is in principle, aspatial. The same is true for its mechanism of implementation, indirectly through market prices and directly through subsidies. As a consequence, the geographical configuration of CAP spending is determined through a bottom-up procedure influenced by the composition of agricultural production, the structure of farms and the structure of agricultural prices. It is not possible to predetermine the distribution of CAP financial resources, either by taking other criteria into account or by implementing a deliberate scheme of a top-down nature. This means that while CAP involves a significant redistribution of income between regions and social groups, this occurs irrespective of existing regional disparities. Though a spatial dimension has recently been introduced to agricultural policy objectives (reform of CAP in 1992, Fifth Environmental Action Plan, proposals of Agenda 2000), this is mainly related to the protection of the environment ("greening of CAP"); regional development issues remain virtually absent.

However, this policy has a potentially very important regional impact, for two reasons. First, because of the very magnitude of the financial resources which are involved. Its share in the EU budget has already been mentioned. In absolute terms, an indicator is the 1995 CAP expenditure, which amounted to 39,947 MECU (36,972 of which concerns the Guarantee section), a sum correspondent to 0.6% of EU GDP (EC, 1995).

The second reason lies in the fact that the recipient of these resources is by definition the agricultural sector. Since the geographical distribution of agricultural population is not homogeneous, and agriculture is the least productive sector of the economy[1]the reduction in regional disparities as a by-product of CAP expenditure, through an income-effect, should be considered as probable. Yet, such an impact cannot be taken for granted, because of both the complexity of CAP allocative mechanism and the almost total lack of institutional (that is, inherent in the official regulations and procedures) co-ordination with regional policy.

The regional impact of CAP has not been sufficiently studied. At the EURRN European Conference "Regional Frontiers" (1997) only one relevant paper was presented (Economou, 1997a), while in the main scientific journals recent articles on the subject are rare. Analyses of the matter can be found in a number of reports and EU services publications, but are mostly brief and sometimes lacking any empirical basis. For instance, the study by Tardifi (1995) on the impact of agricultural price policy on EU cohesion is based on data covering only 1994 and the geographical resolution is that of NUTS2, which is a very aggregate one. Nevertheless, at the international level – that is, between Member States – there is evidence that CAP has been instrumental in the narrowing of disparities in income per head that has actually been observed, especially as far as the "cohesion countries" (Spain, Portugal, Greece and Ireland) are concerned (EC, 1994b, EU, 1995, Mur, 1996, Armstrong, 1995). It is noteworthy that three of these countries have been net beneficiaries from CAP since the 1992 reform (and two were beforehand) (EU, 1996: 5, 62)[2].

Although it is argued that this inter-state effect could be changed[3], it is at the interregional level that the impact of CAP on spatial disparities remains a more open issue. The view that it is possible to discern a positive effect of CAP on the distribution of income between regions in each Member State individually (EU 1996) is not, universally accepted. Indeed, Hadjimichalis (1987) and more recently Dunford and Kafkalas (1992) have claimed that CAP tends to widen existing regional inequalities, by favouring the richer rural areas. The fact that over the past decade regional income inequalities measured on the national context widened in almost all Member States (EU, 1996), despite a reduction in the inter-state disparities, is indeed an indication that the matter is complex and cannot easily be assessed. The recommendation for better co-ordination between regional and agricultural policy in the European Spatial Development Plan (ESDP, 1997) also reflects the climate of apprehension surrounding the issue.

The aim of this paper is to examine some of the aforementioned issues, on the basis of information concerning the expenditure of CAP-Guarantee in Greece over the period 1989 to 1996[4]. The selection of this period is mainly due to the availability and reliability of data; it must also be noted, however, that the beginning of this period is significant, since it coincides with the intensification of Community regional intervention and a new era of structural funds programming.

The Regional Distribution of CAP Transfers in Greece

Agriculture has a particular importance in Greece, since it provides employment to about one fifth of the population and yields some 15% of the Agricultural GDP. The participation rate in the sector is the highest in the European Union, employment's share of the total workforce is about four times the European average. Also, because of the widely dispersed rural population, the income effect of CAP is diffused over a very large part of the national territory.

Greece entered the European Community in 1981 and, since then, has received very substantial sums of money under CAP (Figure 2.1)[5]. Until the middle of the current decade, CAP support which had constantly been augmented, stabilised around 2.7-2.9 Billion ECU (about 7-8% of the total EAGGF expenditure) (LTAPIC, 1998). According to recent estimates, European financial support to local agricultural products annually covers around 40-45% of farmers' net income.

Therefore, an important proportion of the Greek population experienced a significant rise in their level of income. Greece is the sole Member State with a steady long-term increase in the total income from agriculture. Moreover, between 1981 and 1992 the real income per annual work unit in agriculture were augmented by 25%, the corresponding average increase in EU being about 8%. In the same time, the ratio of the average rural to the average urban household income has shifted from 80% to 90%. This is a major economic, social and spatial development, changing long-standing features of city-countryside relations in Greece[6]. These developments – which took place despite the final agricultural output in Greece[7] being less than the European average – are partly due to a reduction in the agricultural active population, but are also a direct result of CAP financial transfers (MNE, 1994, LTAPIC, 1998).

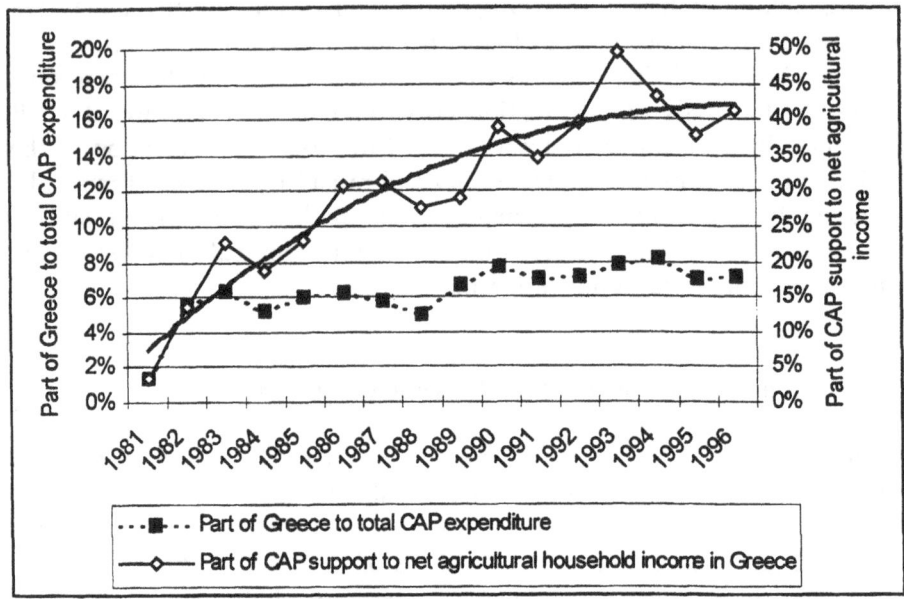

Source: LTAPIC 1998 (Table 4), Ministry of Agriculture 1996

Figure 2.1 The importance of CAP (EAGGF-Guarantee) transfers in Greece

The magnitude of Community resources channelled through CAP (EAGGF-Guarantee), on the one hand, and regional policy (programmes financed by the Structural Funds[8]) on the other, over the examined period, is presented in Figure 2.2 below.

The relation between CAP and structural intervention spending is 44 to 56. Because of the concentration of Objective 1 Structural Funding the relative proportions are different compared to the overall Community Budget, it still represents a very considerable amount, both in absolute and relative terms (as part of the Greek GDP the annual CAP transfers amount to about 5 %). Over the course of the last ten years, receipts from CAP are roughly equal to a "parallel" and unplanned CSF. This means that the possible effect of CAP interventions on regional organisation is potentially very high.

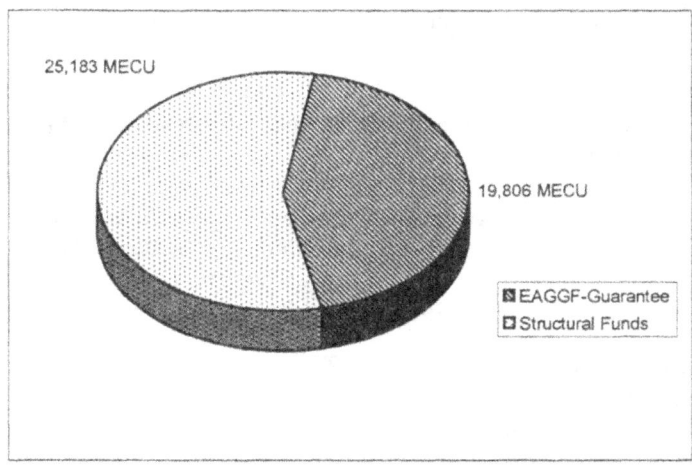

Figure 2.2 CAP and regional policy spending in Greece, 1989-96

The analysis that follows focuses on this question. The geographical level of the analysis is not that of NUTS 2 (regions in Greece) but a more refined one, based on the 53 departments (nomos), a NUTS 3 administrative division of the national territory[9]. This choice gives a more accurate image of the phenomena under investigation and also permits a more significant statistical analysis.

Data regarding the departmental distribution of CAP spending and some other useful indicators are presented in Appendix 1. Table 2.1, summarises some basic statistics.

On the basis of this table, the following observations can be made:

- The inter-departmental differentiation of CAP transfer per head (total population) is medium, with a coefficient of variation (CV) of 0.66 (0.67, excluding Attica). In fact, this variation is less than that of the agricultural workforce as part of the population (0.72). These findings somewhat reduce the probability of a very intense impact of CAP transfer on inter-departmental disparities, which was predicted on the basis of its sheer magnitude. However, a notable impact remains a possibility, depending upon the details of its spatial distribution.
- The variation of CAP transfer per person employed in agriculture is notably higher: 2.17; even excluding Attica, which is an atypical case[10], the variation is 0.95. This means that the spatial differentiation of CAP spending is not a linear function of the differentiation of the place of agriculture in the departmental economy. Consequently, an

asymmetric effect of CAP upon the pre-existing situation, concerning both agriculture and the overall departmental economy, is probable.

Table 2.1   Basic statistics regarding the distribution of CAP transfers at the departmental level

	Average	Standard deviation (sd)	Coefficient of Variation (CV)
CAP transfers, 1989-96, KECU	386,796	844,190	2.18
CAP transfers per head of the total population, 1989-96, KECU	1.92	1.27	0.66
CAP transfers per head of the total population (excluding Attica), 1989-96, KECU	1.93	1.28	0.67
Workforce in agriculture as part of total population 1991	16%	12%	0.72
CAP transfers per person employed in agriculture, 1989-96, KECU	20.22	43.88	2.17
CAP transfers per person employed in agriculture (excluding Attica), 1989-96, KECU	14.38	13.63	0.95

Source: See Appendix 1.

Table 2.2. It contains the correlation coefficients between CAP expenditure (expressed in different ways) and a number of relevant variables, (allows a further investigation of these issues). The main findings are as follows:

- The relationship between the share of a departments total CAP transfer during the 1989-96 period, and the total (national) workforce in agriculture in 1991, is statistically significant, but very feeble. This as is indicated by the coefficient of determination r^2. No more than 6% of the total variation of CAP transfer is explained by the variation in agricultural employment. Furthermore, CAP transfer per head has an equally weak relationship with the workforce in agriculture (r^2=7%). This corroborates the previous comment regarding the variation of CAP transfer per person employed in agriculture. On the whole, these findings imply that the total CAP assistance received by a department is not substantially influenced by the degree of its specialisation in agriculture.

Table 2.2   Correlation between CAP transfers and relevant variables (sums in KECU, departmental level)

Variable 1	Variable 2	r	r^2	t-ratio	Significant at 0.01 level*
CAP transfers 1989-96, KECU	Workforce in agriculture as part of total population, 1991	0.24	6%	-1.7404	yes
CAP transfers per head (total population), 1989-96, KECU	Workforce in agriculture as part of total population, 1991	0.27	7%	1.9874	yes
CAP transfers 1989-96 per employed in agriculture, KECU	Agricultural GDP per employed in agriculture, 1991	0.54	29%	4.4752	yes
CAP transfers per head (total population), 1989-96, KECU	GDP per head, 1989	0.29	9%	2.1542	yes
CAP transfers per head (total population), 1989-96, KECU	Regional policy spending per head, 1989-96, KECU	0.13	2%	0.8943	no
Regional policy spending per head, 1989-96, KECU	GDP per head, 1989	0.22	5%	1.5432	yes

r = correlation coefficient
r^2 = determination coefficient
* The Student t is 1.2991 at the 0.01 significance level.

- There exists a rather high *positive* correlation (r=0.54, r^2=29%) between CAP spending per person employed in agriculture over the period examined, and the agricultural GDP per person employed in agriculture in the department at around the beginning of the period. From the point of view of the objectives of this policy, it is an unexpected situation.   Although at the national level CAP had an overall beneficial effect upon the income of the agricultural population, at a more detailed level of analysis it seems that this effect tends to operate in favour of those already well-off agricultural sub-groups and areas, and strengthens the intra-agricultural disparities.

- There is a *positive* albeit weak correlation between CAP transfers during 1989-1996 and GDP per head in the first year of this period (1989), as is indicated by an r of 29% ($r^2=9\%$). This means that over this period, CAP assistance was more concentrated, in relative terms, in those areas that were already characterised by a higher level of growth. This fact points not only to a lack of synergy, but even to a possible contradiction between agricultural and regional policies, at least at the interregional (departmental) level[11]. It seems that there is some empirical evidence for the disquiet concerning the impact of CAP upon spatial disparities.

- The correlation between a departmental distribution of CAP transfer and regional policy spending is very low and statistically non-significant. This underpins the lack of coherence between agricultural and regional policy and adds further to the impression that, taken independently, the spending of either policy has a positive coefficient of correlation with GDP per head (lines 4 and 6, Table 2.2). It follows that, actual spending dictated by both structural and agricultural policies, tends to be in conflict with the objective of reduction of spatial disparities, by focusing on departments with an already above average position. But its policy has a bias toward different departments, hence the lack of correlation between them.

Figure 2.3 gives a visual expression of the phenomena described above; the names of the departments are not indicated, but they are ranged according to their GDP per head in 1989 (in decreasing order, from left to right). Overall, the graph visually confirms the lack of spatial co-ordination between the two policies suggested by the regression analysis. Given the importance of EAGGF-Guarantee financial flows, this limitation clearly involves a risk that their spatial effects will be inconsistent or even contradictory.

A detailed examination still reveals some interesting patterns. The overall trend of CAP transfer is almost linear, with a slight slope in favour of the richer departments. In comparison, the trend of regional policy transfer is more complex. As a consequence, the aggregate impact of the two policies differs in relation to the level of growth within the departments:

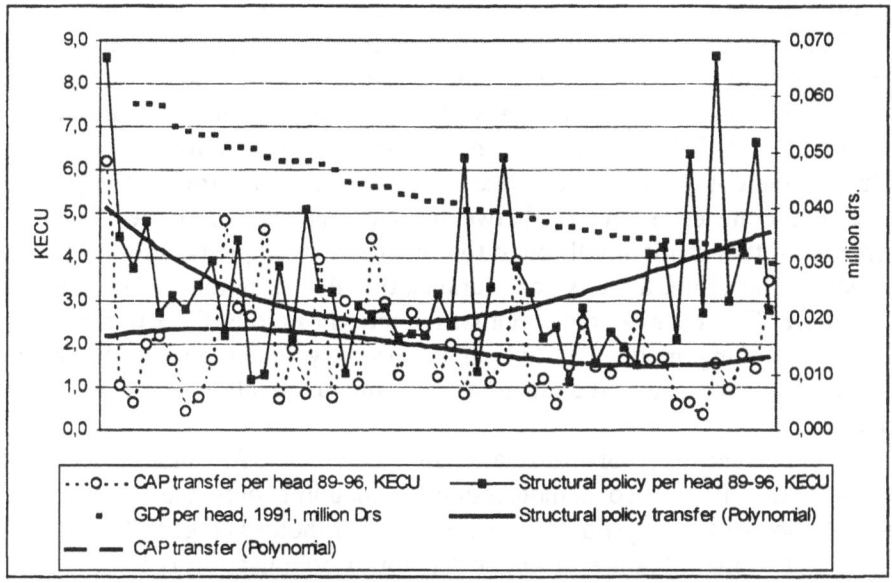

Figure 2.3 CAP and regional policy transfers per head per department

- The departments situated in the upper quarter of the level of development scale are favoured by regional policy but not by CAP;
- Some departments situated in the middle part of the scale receive less than the average from regional policy, but this is often counterbalanced by a higher than average CAP spending. As a consequence, the aggregate EU expenditure in this category is similar, if slightly worse on the whole and with marked individual fluctuations, compared to that of the first category
- The category with a worst handicap, receiving less than the average from both policies, is that of the low-middle growth level departments;
- The departments at the lowest end of the growth scale are clearly disadvantaged from the point of view of CAP assistance, but in their case the regional policy expenditure is quite high. As a consequence, the aggregate spending of the two policies is similar to that of the upper-middle level departments.

The observations above are interesting, since they give a more sanguine image of the complementarity between the two policies, which, in many cases, exhibit some kind of mutually corrective action. This phenomenon should not be dismissed, especially given the fact that regional policy spending is in its self not very well tuned to the existing spatial disparities. Nevertheless, this complementarity is partial and relative. In the first place, the problems underlined during the discussion of the findings in Table 2.2, though tempered by detailed analysis, still remain valid. Secondly, in the cases of the departments with low-middle level growth, or with a low agricultural GDP per person employed in agriculture ratio, the detailed analysis has not modified the original conclusions. Thirdly, the complementarities that have been revealed are not the product of a deliberate co-ordination between CAP and regional policy but have emerged rather randomly through the independent implementation of these policies. Therefore, there is no assurance that this situation will continue in the future. For instance, the eventual reduction of either CAP or Structural Funds financing in the EU, which is currently under discussion, would almost certainly lead to the disturbance of this incidentally achieved partial balance.

The intricacy of the relationship between CAP and geographical growth patterns is also demonstrated by mapping both the departmental distribution of CAP spending overlain by other themes (Figure 2.4).

The main feature of Greek territorial organisation is the existence of an S-shaped growth axis, passing though Athens and Thessaloniki, the two Greek metropolises, and extending from Patra to Kavala (Economou 1993). It can be observed that, in general, the highest CAP spending per head is located along the axis. Peripheral departments are handicapped, even when they are predominantly rural (i.e. with a high percentage of agricultural workforce), while departments on the axis receive relatively high amounts of CAP spending, despite their generally differentiated and mainly non-agricultural economic base. This spatial selectivity of CAP is linked with the fact that its support of different products is very unequal[12], the most favoured ones being those produced in larger quantities in the lowlands located (in large part) along the axis (LTAPIC, 1998). It is also significant that the agriculture holdings declaring a taxable income – presumably the larger ones – are concentrated in the same zone (Sivignon, 1996). This selective pattern should be superimposed on those trends identified previously. Overall, it accentuates the CAP bias in favour of the more developed areas.

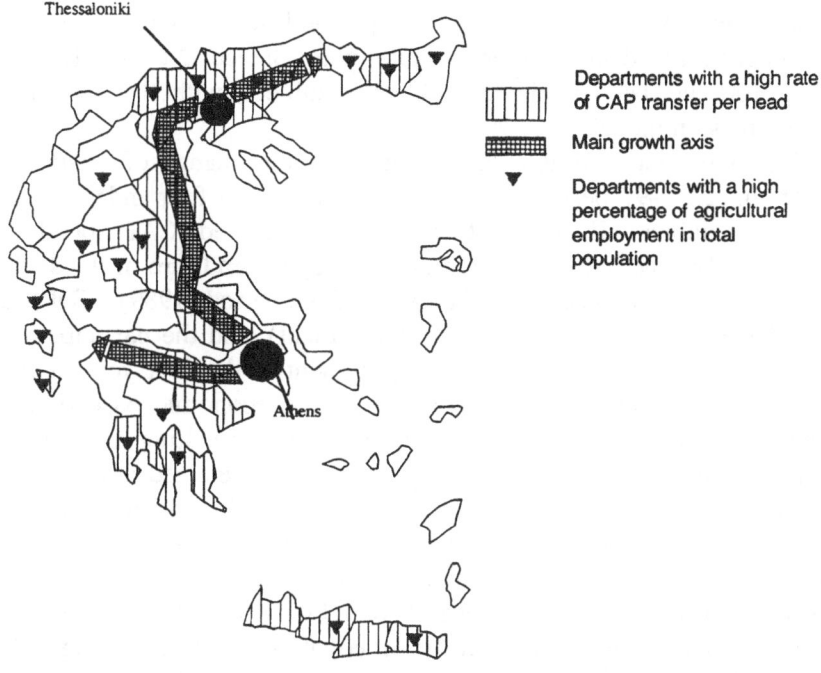

Note: High rate of CAP transfers per head and high percentage of agricultural employment = more than the average of the respective interdepartmental distributions (1.84 KECU and 16.5%)

Figure 2.4 Distribution of CAP transfers, specialisation on agriculture and position in relation to the main growth axis in Greece

The Regional Impact of CAP

The analysis so far refers to the spatial distribution of CAP spending. Obviously, there is a strong link between financial spending and growth patterns. On the other hand, the former is not a direct result of the latter, since the final outcome is determined by other factors, such as efficiency in the use of mobilised resources. This is especially relevant in the present case, since it is well-known that the EAGGF-Guarantee resources in Greece were used exclusively to raise living standards in rural areas and not for the financing of investments (Maraveyas, 1989 and LTAPIC, 1998).

To produce a definite answer to the question of the *actual* impact exerted by CAP expenditure upon regional structures necessitates, therefore, specific research focused upon growth rates. However, such an endeavour is very difficult without the use of econometric models, for which the necessary information (input-output tables, time-series, detailed GDP cross-sectional data) is largely lacking in Greece. Nevertheless, a tentative effort has been made to tackle the subject, with the use of a multiple regression model.

The dependent variable in the model is the change of GDP per head over the 1989-97 period[13]. Initially, the following four independent explanatory variables, all at departmental level are used, CAP spending during the 1989-96 period (CAP), regional policy expenditure in the same period (CSF), GDP per head in the first year of the period (1989) (GDP) (a measure of the initial dynamism of the department), and the participation of the agricultural workforce in total employment at the first year of the period) (as an indication of the structure of the economic base of the department) (AGR). However, the F ratio of the resulting regression equation was not enough for the null hypothesis to be rejected[14]. The same holds true for the equation arrived at using only the first three variables above[15]. Since the number of observation is not very low (51), these results suggest either inaccurate data (which is, to some extent, a plausible hypothesis...) or an incoherent pattern of spatial growth, or both.

In the end, a statistically significant equation was found using only two independent variables, both related to the European Union funding. The regression equation is as follows:

$$y=-0.06+0.01CAP-0.03CSF \qquad\qquad (3)$$

The multiple R^2 is 0.35, with an F ratio of 3.408, higher than the F statistic for the 0.05 level of significance, which is 3.22.

Still, the regression remains unsatisfactory. CAP spending (CAP) has a positive but very low coefficient which suggests no more than trifling effect upon growth rate and, hence, regional disparities. This means that the explanatory force of this variable is quite weak. Nevertheless, such a result is compatible with both the use of CAP aid almost exclusively for individual consumption, and the characteristics of the spatial distribution of CAP, which has been analysed above. The former factor explains why, apart from an immediate increase in revenue, CAP spending did not have longer-term multiplying effect. The latter factor, that is, the lack of strong links between CAP spending and the internal economic structure of the departmental economy, repeatedly referred to throughout this paper, obviously justifies the lack of a systematic effect upon economic evolution. Interpretations based on other approaches could be added to these. The suggestion made by Sarris and Zografakis (1995) that, in Greece, current forms of farm support might have worsened the farm adjustment problem implies that high levels of spending are compatible with low or erratic structural impact. These arguments are specifically related to Greek conditions, but some more general criticism of CAP could be added as well. The claim made by Alvarez-Coque (1996) that CAP payments may be blocking structural adjustment and maintaining income inequalities in the rural sector, is just one among those which ascribe to CAP a long-term anti-cohesive effect because of its contribution to conserving a high percentage of agricultural employment, which is presumably less productive.

As for the other independent variable, regional policy expenditure, its coefficient in the equation is negative, denoting an analogous negative effect made by this variable on the growth pattern! Since the *raison d' être* of the Structural Funds is the promotion of the growth of those regions which lag behind, this finding is especially shocking. To some extent, this result may be attributed to inadequate data. Indeed, the information regarding the spatial distribution of the Structural Funds spending, derives from a study programme carried out by the Ministry of the Environment, Planning and Public Works which is the first of its kind in Greece. This study is generally considered as a useful first step which, however, necessitates sustained follow-up in order to produce more reliable results. Nevertheless, apart from this factor, it is a well documented fact that the effectiveness and efficiency of both First and Second Community Support Frameworks in Greece are very limited[16]. Under these circumstances, the lack of a quantifiable relationship between the growth process and the European regional policy expenditure becomes a more convincing explanation – although this does not render the very fact more justifiable.

Conclusion

To summarise, the main conclusions of the paper are as follows:
- CAP spending is a very important part of EU expenditure, from a quantitative point of view. Consequently, the investigation of its influence on growth and regional development is very relevant
- The inter-departmental distribution of CAP transfers is hardly a function of the relative weight of agriculture in the local economy, while at the same time, it shows a bias in favour of the more dynamic components of the agricultural sector.
- The overall synergy between CAP and regional policy spending is limited. A more detailed analysis reveals a certain amount of

complementarity between the two policies, but this outcome is non-deliberate, partial and unstable.
- The eventual impact of CAP transfers on spatial development disparities seems negligible.

In view of the magnitude of CAP resources, these conclusions have serious consequences; at least if the experience in Greece during the 1989-96 period is not atypical and is to be found in other EU countries. In this case, even if CAP is also related to other objectives (social, environmental...), the preservation of such a costly policy without positive effects on either economic competitiveness or spatial cohesion has to be reconsidered and serious changes are needed.

At least two lines of action seem advisable. First, the role of (low) income as a factor of EAGGF-Guarantee assistance allocation must be enhanced[17]. The reduction of price support and increased focus on direct payment, initiated in 1992 and further adopted in the proposed new reform of CAP under Agenda 2000 (Fischler, 1997a), points effectively in this direction. However, the difficulties in adapting to some other aspects of the proposed reform (market orientation, multi-functionality...) experienced by the more backward agricultural economies, should also be taken into account[18]. Otherwise, the new CAP could lead to an accelerated internal polarisation of the European agricultural sector.

Secondly, a substantial increase in the co-ordination between agricultural and regional policy is urgently required. Again, this is included in the goals of the reform of CAP under Agenda 2000 (Fischler, 1997b), but its operational implementation will impose major changes in the regulations and procedures of EAGGF-Guarantee. An obvious necessity is that, not only the income of the particular farmers, but also the average income per head of the total population in an area, should be considered. In practical terms, this could necessitate a regionally differentiated allocation mechanism, instead of one based merely on sectoral or product-market considerations. Such a direct link between cohesion and agricultural policy could, in fact, facilitate the shift to direct payment, which, in general, tends to be politically less acceptable than indirect support through price policies (Binswanger and Deininger, 1997: 1976-1977). In a different approach, a measure that could be advocated is a substantial modification of the balance between the Guarantee and Guidance sections of EAGGF in favour of the latter[19]. Although the subject is complicated, the Greek experience, at least, points clearly in this direction, because of the overwhelming proportion of CAP aid which was used for the promotion of income and not for the increase of investment.

Notes

1   Although this does not necessarily mean that income in the agricultural sector is always very low. In fact, for the years for which a comparable set of data was available, the income of the average farm household exceeded that of the average of all households in all member states apart from Portugal (Jovanoviæ, 1997: 116). Despite the fact that such a comparison of averages may be misleading, especially in this case, this evidence suggests strong intra-agricultural differentiation.

2   Which led to the increase in the support to weaker regions and countries with a greater workforce in the agricultural sector (ESDP, 1997: 39).

3   In the European Spatial Development Perspective it is estimated that, in the future, areas with less intensive producers could be handicapped in relative terms by the proposals of Agenda 2000 for CAP reform (ESDP, 1997: 40).

4   CAP spatial impact is multifaceted. This paper focuses on the aspects which are related to the transfer of resources to farmers through EAGGF-Guarantee, but this does not mean that other issues as, for instance, the influence on living standards through the effect on prices or the environmental dimension of this policy are negligible. For a discussion of the latter, see Economou (1993).

5   CAP transfer includes both direct and indirect spending by EAFGG-Guarantee. The calculations (both for the aggregate sums and their geographical distribution) have been made by the Ministry of Agriculture, following a request from the Ministry of the Environment, Planning and Public Words. Part of this data has been used in the latter's recent study programme "Spatial impact of Community programmes and policies", but this is the first time that data have been publicly presented and analysed.

6   At least until the middle-70s, the urban-rural income gap in Greece tended to be increasing (Vergopoulos, 1975). Such a trend is still visible in many developing non-European countries (Binswanger and Deininger, 1997).

7   Final agricultural output in Greece augmented by 6.5% between 1980-1990 compared to a Community average of 15.2% (Agricultural Situation in the Community, 1991: T/33).

8   Including all IMFs, First and Second CSFs, and Community Initiatives, all financed by Structural Funds (ERDF, ESF, EAGGF-Guidance) and, moreover, projects which receive assistance from the Cohesion Fund. Although the latter is not technically a structural Fund, it is a component of EU regional policy.

9   In the above-mentioned paper by Economou (1997a), the spatial resolution of data and analysis was more aggregate, referring to the 13 Greek regions (NUTS2).

10 Attica is simultaneously a department and a metropolitan region that virtually coincides with greater Athens. It is very different from the other departments, and contains some 40% of the total population. In terms of CAP transfers per person employed in agriculture, Attica's rate is very high, about 15.5 times greater than the average. This peculiarity is attributable to the very low percentage of agriculture in the total workforce (about 1%), as well as to the concentration in the city of a relatively great number of the official seats of agri-industrial enterprises which are beneficiaries of CAP subventions.

11 At the level of the 13 NUTS2 regions (as opposed to the 50 NUTS1 departments) the correlation between the same two variables is almost null and, in any case, statistically non-significant (Economou, 1997a).

12   In recent years, cotton, cereals, tobacco, olive oil, and vegetables have accounted for about 85% of total CAP transfers (LTAPIC, 1998.).

13   Despite its shortcomings, GDP is a convenient measure of the level of development, and the main goal of the EU regional policy for those areas which lag behind (Objective 1) [areas with a GDP per head equal to or less than 75% of the Community average] is explicitly related to it.

14   The resulting equation in this case is

$$y=-0.12+0.02CAP-0.02CSF-1.03GDP-0.18AGR \qquad (1)$$

The value of the F ratio is 1.977 while the F statistic is 2.565 for the 0.05 level of significance with (4, 51) degrees of freedom.

15   The resulting equation in this case is

$$y=-0.09+0.02CAP-0.02CSF-1.05GDP \qquad (2)$$

The value of F ratio is 2.313 while the F statistic is 2.880 for the 0.05 level of significance with (4, 51) degrees of freedom.

16   See Economou (1997b), regarding the low effectiveness of the First CSF. The first Evaluation Report for the Second CSF (KEPE-REMACO 1998) indicates that, despite a long-term amelioration, the growth impact of the programme until 1996 (that is the period covered by the present paper) was marginal, due to a very low absorption rate of the earmarked funds.

17   Jovanoviæ (1997: 127) makes analogous propositions, but in a different context (mainly an attempt to reduce of CAP cost and consumer prices).

18   See a number of papers presented in the ESDP Seminar "For a new rural-urban partnership" in Salamanca (Spain, 1998), for instance Sallard (1998) and Saraceno (1998).

19   Such a measure could also counter-balance a tendency of redirection of existing structural spending from rural to urban areas, linked to the fact that regional problems are starting to concentrate in the urban areas. See, for instance, UE (1997).

References

Armstrong, H. (1995), 'Convergence among regions of the European Union, 1950-1990', *Papers in Regional Science*, vol. 74, No. 2, pp. 143-152.

Ashim, B. and Dunford, M. (1997), 'Regional Futures', *Regional Studies*, vol. 31, No. 5, pp. 445-455.

Binswanger, H. and Deininger, K. (1997), 'Explaining Agricultural and Agrarian Policies in Developing Countries', *Journal of Economic Literature*, vol. XXXV, pp. 1958-2005.

Blalock, H. M. (1979), *Social Statistics*. McGraw-Hill Kogakusha: Tokyo.

Dunford, M. and Kafkalas, G. (1992), 'The global-local interplay, corporate geographies and spatial development strategies in Europe', in: Dunford M. and Kafkalas G. *Cities and Regions in the New Europe*. Belhaven Press: London, pp. 3-38.

EC (European Commission-Directorate General for Economic and Financial Affair) (1994a), *EC Agricultural Policy for the 21st Century*. EC: Luxembourg.

EC (European Commission) (1994b), *Competitiveness and cohesion: trends in the regions*. EC: Luxembourg.

EC (European Commission) (1995), *The Agricultural Situation in the Community, 1994 Report*. EC: Luxembourg.

Economou, D. (1993), 'New forms of geographical inequalities and spatial problems in Greece', *Environment and Planning D: Society and Space*, vol. 11, No. 5, pp. 583-598.

Economou, D. (1997a), 'The regional impact of the Community Agricultural Policy. The case of Greece'. Paper presented in the European Urban and Regional Research Network (EURRN) 1997 European Conference, 20-23 September 1997, Europa-Universitat Viadrina Frankfurt (Oder), Germany.

Economou, D. (1997b), 'The impact of the First Community Support Framework for Greece: The anatomy of a failure'. *European Urban and Regional Studies*, vol. 4, No. 1, pp. 71-76.

ESDP (1997), *European Spatial Development Perspective*. Presented at the informal meeting of Ministers responsible for spatial planning of the Member States of the European Union, Noordwijk, June 1997.

EU (European Union-Regional Policy and Cohesion) (1995), *Cohesion and the development challenge facing the lagging regions*. EU: Luxembourg.

EU (European Union) (1996), *First Report on Economic and Social Cohesion, 1996. Preliminary edition*. EU: Luxembourg.

Eurostat (1993), *Farm Structure. 1993, Survey*. EC: Luxembourg.

Fischler, F. (1997a), 'Agenda 2000 - reform of the EU's agricultural policy'. Speech, to the meeting of COPA/COGECA, Brussels, 14 November 1997.

Fischler, F. (1997b), 'Agenda 2000: CAP reform proposals'. Speech, to the Council of Ministers of Agriculture, Echternach, 7-9 September 1997.

García Alvarez-Coque, J. M. (1996), *Agri-food policy in member states: Producers' perspective*, School on European Integration: Rome.

Hadjimichalis, C. (1987), *Uneven Development and Regionalism. State, Territory and Class in Southern Europe*, Croom Helm: London.

Jovanović, M. (1997), *European Economic Integration*, Routledge: London.

KEPE-REMACO 1998, *First Report of the Evaluation the CSF 1994-1999 for Greece*, Athens.

LTAPIC (Log-term Agricultural Policy Investigation Committee) (1998), *Competitive agriculture and rural development*. National Bank of Greece: Athens.

Maraveyas, N. (1989), *The accession of Greece to the EC: The effects on the agricultural sector*, IME: Athens (in Greek).

Ministry of Agriculture (1996), *Management of OP 'Development of Agriculture 1994-1999'. First report*. MA: Athens (in Greek).

MNA (Ministry of National Economy) (1994), *Regional Development Plan of Greece 1994-1999*, MNA: Athens (in Greek).

Mur, J. (1996), 'A Future for Europe? Results with a Regional Prediction Model'. *Regional Studies*, vol. 30, No. 6, pp. 549-565.

Sallard, O. (1998), 'Making rural assets a competitive advantage', Paper presented in the ESDP Seminar For a new rural-urban partnership. 15-16 Oct. 1998, Salamanka, Spain.

Samothrakis, V. (1997), 'Trends of specialisation and concentration of the agricultural output in Greek regions', *Economical Annuals*, May-June 1997 (in Greek), pp. 12-27.

Saraceno, E. (1998), 'Changing rural-urban relationships: an opportunity for both rural and urban areas', Paper presented in the ESDP Seminar For a new rural-urban partnership, 15-16 Oct. 1998, Salamanka, Spain.

Sarris, A. H. and Zografakis, S. (1995), *Agricultural income statistics and policy: A view from Southern Europe*, Department of Economics, University of Athens: Athens.

Sivignon, M. (1996), 'Les métamorphoses des campagnes grecques', *Méditerranée*, vol. 1, No. 2, pp. 79-86.

Tarditi, S. (1995), *Impact of the agricultural price policy on EU cohesion-a regional analysis*. Research report, DG XVI: Brussels.

UE (Union Européenne) (1997), *L' Europe des villes*, UE: Luxembourg.

Vergopoulos, C. (1975), *The rural question in Greece*, Themelio: Athens (in Greek).

Appendix 1   The distribution of CAP funding among the Greek departments ("nomos")

	CAP transfers, 1989-96, KECU, current prices		Structural Funds aid, 1989-96, current prices		GDP per head, KECU, Million Drs, 1970 prices	
	Total	Per head	Total	Per head	1989	1996
Department	(1)	(2)	(3)	(4)	(5)	(6)
Yannina	96,636	0.617	332677	2.1	0.034	0.034
Arta	139,692	1.769	332009	4.2	0.032	0.029
Thesprotia	67,223	1.536	381501	8.6	0.033	0.031
Preveza	95,114	1.633	368434	6.3	0.039	0.032
Larisa	790,681	2.947	767300	2.8	0.044	0.041
Magnisia	141,464	0.721	749102	3.8	0.048	0.052
Trikala	225,004	1.627	565224	4.1	0.034	0.030
Karditsa	494,393	3.904	481151	3.8	0.039	0.048
Kerkyra	136,187	1.280	230070	2.1	0.042	0.047
Lefkada	20,428	0.962	63313	3.0	0.032	0.032
Kefallinia	39,291	1.216	101842	3.1	0.041	0.038
Zakynthos	142,195	4.417	86237	2.6	0.044	0.038
Aetoloakarnania	330,490	1.456	351248	1.5	0.036	0.043
Achaia	315,289	1.063	859512	2.9	0.044	0.044
Illia	258,452	1.463	199291	1.1	0.037	0.037
Fthiotida	479,969	2.824	747750	4.4	0.051	0.038
Evrytania	15,193	0.618	154691	6.4	0.034	0.027
Fokida	37,481	0.848	225532	5.1	0.048	0.042
Viotia	815,986	6.198	1697711	12.7	0.086	0.076
Evia	133,539	0.650	782536	3.8	0.059	0.046
Korinthia	279,667	2.011	682013	4.8	0.059	0.062
Argolida	382,371	3.942	320459	3.3	0.048	0.044
Arkadia	80,696	0.763	336439	3.2	0.047	0.044
Messinia	396,490	2.389	363287	2.2	0.041	0.038
Lakonia	238,110	2.496	269308	2.8	0.036	0.033
Lesvos	170,878	1.626	201028	1.9	0.035	0.035
Chios	18,909	0.365	141817	2.7	0.033	0.028
Samos	37,593	0.900	134314	3.2	0.038	0.035
Cyclades	42,269	0.453	262492	2.8	0.054	0.043
Dodekanisa	124,616	0.775	547764	3.4	0.053	0.054
Chania	263,715	1.988	326114	2.4	0.041	0.051
Rethymno	187,442	2.716	156875	2.2	0.042	0.046
Irakleio	783,490	2.992	349708	1.3	0.045	0.055
Lasithi	133,135	1.872	150348	2.1	0.048	0.052
Attiki	5,685,870	1.623	10876492	3.1	0.054	0.057

	CAP transfers, 1989-96, KECU, current prices		Structural Funds aid, 1989-96, KECU, current prices		GDP per head, Million Drs, 1970 prices	
	Total	Per head	Total	Per head	1989	1996
Evros	243,587	1.685	607485	4.2	0.034	0.043
Rodopi	359,038	3.454	287494	2.8	0.030	0.028
Xanthi	120,890	1.332	205296	2.3	0.035	0.038
Drama	115,505	1.199	205756	2.1	0.038	0.049
Kavala	297,309	2.188	365767	2.7	0.058	0.053
Serres	510,334	2.638	295103	1.5	0.034	0.042
Thessaloniki	2,443,721	2.610	1109696	1.2	0.050	0.052
Chalkidiki	147,085	1.630	358913	3.9	0.053	0.049
Kilki	182,171	2.229	109110	1.3	0.040	0.055
Pella	634,434	4.601	177444	1.3	0.049	0.044
Imathia	674,322	4.848	305871	2.2	0.051	0.050
Pieria	98,131	0.851	733421	6.3	0.040	0.042
Florina	59,472	1.121	176114	3.3	0.039	0.038
Kozani	155,623	1.035	670775	4.4	0.071	0.073
Kastoria	32,049	0.607	126204	2.4	0.037	0.024
Greveno	53,008	1.442	245077	6.7	0.030	0.027
TOTAL	19,726,639	1.937	30575115	3.0	0.048	0.050

Data: (1), (2): Ministry of Agriculture (unpublished data, (3), (4): Ministry of the Environment, Planning and Public Works, Regional studies of the impact of Community Programmes and Policies, (5), (6), Ministry of National Economy-Centre of Economic Planning and Research (KEPE)

3 The Role of Local Development Policies

IVAN TUROK

Introduction

Wide economic and social disparities within many European cities and regions threaten political stability and general economic progress. Many local and national authorities have begun to respond to localised concentrations of unemployment, poverty and related social problems with a range of novel area-based policies and initiatives. They provide ways of directing development support more closely to the areas and communities with the greatest needs than traditional regional policies and programmes. The EU Structural Funds have played an important supporting role in some instances and a leading one in several others.

Some initiatives involve a local partnership of public organisations, community groups and the private sector seeking to pursue commonly-agreed regeneration strategies and drawing support from mainstream public policies and programmes. Other initiatives are set up and run by communities themselves, often with external financial support, including community businesses, multi-purpose community centres, voluntary clubs, co-operative food schemes, credit unions and local exchange trading systems (LETS). A third group, are community-based organisations with powers and resources to provide services and encourage regeneration on behalf of and under the influence of the community, but with other stakeholders also involved.

A key objective of most local policies is to alleviate poverty and social inequality by redistributing resources, jobs, training and other opportunities towards disadvantaged communities. Another important aim is to promote local economic development and hopefully contribute to city-wide prosperity by converting underused property and other assets into

35

more productive units, in the process securing additional investment and jobs for the city.

Important questions surround the appropriate composition and scale of local development policies and their relationship to wider regional and national policies. For instance, are they substitutes for wider action or complementary to it? Others questions surround the institutional mechanisms that could facilitate the combined pursuit of social improvement, lasting economic development and popular involvement in the process of regeneration. There is considerable rhetoric and ambiguity around the whole subject, apparent in the terminology of partnerships and pacts, economic and social inclusion, holistic and comprehensive regeneration, community empowerment and capacity building, and labour market adjustment and flexibility.

This chapter adds some reality to the discussion of local development policies and institutions by examining practical experience in three major European cities. Glasgow (Scotland), Dublin (Ireland) and Duisburg (Germany) have been grappling with the challenges of bringing real improvements to the quality of life in poor communities, in some cases dating back over a period of several decades. Their collective experience illustrates some of the ways in which local development can be made to work, as well as the strengths and limitations of a local approach.

Operating in difficult economic circumstances and with an unsympathetic national government, city and regional authorities in Glasgow established a unique network of independent 'local development companies' (LDCs) in areas of greatest need. They have since developed wide-ranging capabilities, often by working closely with other local interests and organisations, and applying energetically for European and other funds. Dublin's priority neighbourhoods also suffer high unemployment, despite Ireland's recent economic growth. Their 'local partnerships' emphasise the 'softer' dimensions of regeneration. They have had greater support from Europe and national government agencies, but less from their local authorities. Duisburg has been exploring an integrated approach to improving areas of severe industrial decline and racial tension through a range of locally based projects, also supported by EU funds. This follows an era dominated by physical renewal organised by the established regional and local authorities.

Arguments For and Against

One of the obvious arguments for the localisation of development policy is to allow support to be targeted to where it is most needed. This could confer wider economic benefits in terms of encouraging growth in non-inflationary ways, bringing unemployed groups closer to the labour market through improved skills and job access initiatives (thereby increasing labour supply), and ultimately reducing public expenditure on the social symptoms and consequences of decline. In addition, local policies may be more efficient than city or regional policies in permitting localised problems to be tackled in more rounded and responsive ways. They could help to co-ordinate and integrate labour demand- and supply-side policy measures, so that, for example, there are appropriate jobs available after people completed training courses. Other economic arguments include the potential for local development organisations to identify and transform vacant land and underused buildings, underperforming businesses and latent entrepreneurial abilities into more productive assets, in the process creating jobs, raising incomes and improving services.

The conventional social arguments for local regeneration policies include the unacceptability of increasing levels and concentrations of poverty, deprivation and others features of social exclusion in otherwise prosperous societies. Poor health, housing, education, high crime, drug misuse, family breakdown, racial tension, dependence on state benefits and other problems also tend to be associated with unemployment and poverty.

Some institutional considerations may also be important. Stronger organisational networks and local capacity may draw in additional public and private resources by designing suitable projects for investment and tailoring economic and social services better to suit local conditions. The ability of communities to resolve local difficulties and promote common interests should be enhanced through greater organisational strength and ability to influence wider resource allocation processes. Beyond this, the growth of local partnerships may encourage more dialogue and interaction between sectional interests, and a more open and inclusive approach to decision-making. This may generate what has been called 'social capital', i.e. long-term commitment to the development of local people and other assets, by involving a range of interested stakeholders in joint problem-solving, policy innovation and transfer of good practice.

There are also some counter-arguments to the localisation of development policy. One is the danger that local development organisations amount to little more than a wasteful distraction and additional layer of bureaucracy with little real power or resources. Local partnerships may get diverted from the fundamental objectives of regeneration by the effort required to develop organisational structures and plans, agree priorities and co-ordinate actions. A UK Government discussion paper recently stated that: "there has been a danger of paper partnerships, set up to secure funding and little else" (DETR, 1997, para. 5.4).

If local organisations prove to be more significant, there may be some concern about the fragmentation and unevenness of economic and social development that results, especially if they compete with each other unhelpfully and unproductively. Local institutions are bound to vary in their effectiveness, leaving some communities potentially much worse off than others in the absence of compensatory regional and national policies, perhaps simply because they started off in a poorer position and could never catch up.

There may be question marks too about whether local initiatives can tackle deep-seated problems associated with wider causes, such as the collapse of traditional industries, pervasive decentralisation of jobs and population from the cities, structural divisions between regional economies, short-termism of the financial system, restrictive macro-economic policies, and inequalities in the labour market. Given the fiscal pressures governments face, they may be attracted to local policies as visible alternatives to national action, which would limit the overall capacity to confront such issues.

An alternative, more positive scenario may involve some combination of complementary national, regional and local actions, i.e. coherent multi-level development strategies. Local initiatives could also be used to experiment with measures that, if successful, get built into mainstream policies, or they could be given the power to adapt and integrate national resource streams according to local circumstances.

A Simple Analytical Framework

Local development encompasses several different objects of policy and institutional configurations. Bearing in mind the danger of oversimplification, Figure 3.1 presents a summary framework to introduce some of the basic differences of emphasis regarding the 'what' and 'how' of area-based measures. One of five broad policy objects and one of two institutional approaches often lead strategies, although these are not the only possibilities. Nor are they pursued in isolation since strategies generally span several categories and may purport to be comprehensive. In practice, there are variations in priority and composition, reflecting local needs, development potential and political preferences. Certain policy choices are also associated more closely with some institutional arrangements than others. The policy composition and institutional responsibilities are bound to change over time as experience develops. The distribution of asterisks in Figure 3.1 simply illustrates in broad terms one possible arrangement at a particular point in time, albeit one that is not uncommon among many current local partnerships. Community-based organisations are more likely to have responsibility for people-oriented functions and tailored services while mainstream agencies tend to retain control over major expenditure programmes, high volume services and perhaps technically difficult functions, although this is a very broad generalisation.

Policy objects	**Institutional approaches** Bending main public sector policies and programmes	Building community-based development organisations
Business development *(job creation)*	*	**
Human resource development *(skills & job access)*	**	***
Physical business infrastructure *(job creation)*	**	*
Neighbourhood improvement *(housing and environment)*	***	**
Social economy *(community services and jobs)*		**

Figure 3.1 Approaches to local development

The distinction between policy objects is fairly straightforward. Business development seeks to promote industrial diversification and modernisation by improving the rate of formation and growth of indigenous enterprises, through the provision of information, technical advice, management training, finance for investment, marketing support and other services. Human resource development aims mainly to increase the local share of jobs in the wider labour market by enhancing people's employability and job access through employment counselling, career guidance, confidence-building, core and vocational skills, work experience and job search support. Physical business infrastructure covers efforts to make places more attractive for productive investment by improving the road network, clearing derelict sites, providing serviced land, building premises and offering incentives to attract incoming employers. This may be essential to deal with the legacy of industrial decline and to upgrade the basic physical condition of old cities.

Neighbourhood development encompasses efforts to retain and attract population by improving the desirability of places to live through refurbishing the housing stock, improving the environment, enhancing leisure and other facilities, and acting on crime and safety concerns. Housing renewal is a long-established focus of area-based policies, but is being added to and sometimes superseded by economic and social programmes addressing the causes of poverty and deprivation. Finally, the concept of the social economy blurs conventional distinctions between economic and social development by supporting multi-purpose activities that have less chance of becoming viable in commercial terms and tend not to be funded by a single source of public subsidy. They often involve socially-useful services that provide alternative employment to the mainstream economy where there is a jobs shortfall, or a route back into the mainstream through work experience, training, personal development, childcare and other support services.

A concern about unemployment is common to all five themes because of its centrality to other local problems. Business and property development seek to increase labour demand (creating jobs in places that need them most), whereas HRD and neighbourhood development are more concerned with labour supply (altering the character, size and geography of the labour force). There is a debate among economists and policy-makers about the relative importance of these methods of reducing unemployment. The answer is bound to vary according to labour market conditions. In situations of high city-wide unemployment, increasing the demand for labour may be the priority, but where unemployment is more spatially confined, supply-side measures may be more effective. As indicated earlier, one of the strengths of local policy approaches is their ability to pursue both demand- and supply-side measures in tandem.

Figure 3.1 also makes a broad distinction between two institutional approaches. The central idea of the first is the redirection of government policies towards marginalised communities through 'bending' mainstream resource streams and perhaps also relaxing certain rules, regulations and occasionally even taxes which are thought to impede local development activity. It may involve some local delivery, refashioning or co-ordination of services such as education, training and housing to meet local needs more effectively than through standard national programmes or specialised functional policies. Locally focused forums or partnerships of city and regional organisations have emerged to pursue this in some places.

This approach has the advantages of ensuring a basic level of government commitment and avoiding some of the financial and bureaucratic diversions of having to build new organisations. However, its impact may be limited by statutory responsibilities restricting the extent of bending and by the lack of ultimate local control and flexibility. Community-based institutions are likely to remain undeveloped and local influence over resource flows restricted. The approach is subject to wider political support and vulnerable to the imposition of policy changes without local consultation. Opposition to an area's special treatment from other areas, or redistribution towards deprived communities generally, may make the policy difficult to sustain.

The second approach involves building organisational capability within localities to bring decision-making closer to community needs and to increase the scope for streamlined, more integrated action. Local organisations may bid for resources from specially created public funds, which are easier to target and devolve than main programmes. Increased local control over resources gives more scope to devise relevant policies without external constraints. Community-based organisations may become catalysts for area improvement by providing technical expertise and negotiation skills to influence wider funding and investment decisions. Key attributes include an enterprising outlook, proficient networking and a mixture of strategic vision with practical competence in project management and financial packaging. These features tend to be missing from the simpler, looser partnership arrangements based on mainstream bending and co-ordination of regional and national organisations.

There is a spectrum within the second approach between total community ownership and control over decision-making and the creation of units that incorporate a slightly wider range of experience and perspectives as well. The former emphasises the virtues of self-reliance and community determination to release local energy, ideas and commitment. It involves a transfer of authority and responsibility to projects and enterprises owned and managed by the community. A DETR paper expressed this rather starkly and over-simply: "Local people need to be more than consulted and involved ... Ultimately if regeneration is not owned by the community, its benefits will not endure" (DETR, 1998, p.2). The latter is based on bringing different stakeholders into the equation with a view to achieving a more substantial impact as a result of drawing in additional organisational and financial resources.

The emergence of this approach may also reflect increasing rivalry between areas for jobs and investment and financial restrictions facing public bodies. At worst, the creation of community-based agencies may mean public authorities limiting their responsibility for tackling difficult problems and providing essentially token support to the poorest areas. At best, there may be a stimulus to substantial activity from bringing different sectional interests together in joint problem-solving and building long-term relationships around a focused agenda. Collective action organised with the constructive objective of promoting economic and social development could have all sorts of benefits for localities.

Glasgow's Experience

Glasgow's experience of area regeneration dates back over three decades. An emphasis on housing renewal in the 1970s was consistent with the city's appalling housing conditions at the time. This broadened out into a concern with social and economic issues in the 1980s, following an extended period of economic decline, rising unemployment and increasing poverty. The city lost 63,000 jobs (one in seven) between 1971 and 1981, and another 54,000 between 1981 and 1996. The loss of manual jobs has been a particular problem - nearly one in three disappeared between 1981 and 1991 alone as a result of deindustrialisation.

Faced with an unsympathetic national government, the city and regional authorities began to respond in several ways. One was to devolve certain powers and resources to the most deprived neighbourhoods and experiment with different consultative arrangements. Other responses took the form of special city-wide economic and social programmes, sometimes oriented towards the poorest communities. Compared with other cities in Britain, substantial initiatives were also developed in relation to community businesses, self-employment, targeted training, work experience and environmental projects of various kinds. From this experience some consensus emerged among the public authorities in the 1990s about several principles for making progress.

One feature is that the city authorities cannot afford to ignore any of the eight sizeable concentrations of poverty in the city, or be overly selective in their support. The extent of deprivation across the city also means that bending mainstream programmes is particularly difficult and may not be very fruitful ultimately. A second theme is that economic development warrants special attention since unemployment underlies many of the other problems and more productive activity is needed in the city to expand employment, increase local income and improve services. A third is that commitment to development needs to be long-term, since the problems of decline and poverty are far-reaching. One way of achieving this is to establish independent local organisations that are managed competently and funded securely. Fourth, regeneration activities cannot rely on redistributing existing resources within the city, but need to draw in additional public and private sector investment, especially against a background of long-term population decline and cuts in public sector funding. Fifth, the community has an important role to play in ensuring local support for regeneration efforts, identifying priorities and shaping service provision as vital users.

Although not formally stated, these views influenced the setting-up and increased support given to local development companies (LDCs) in each of main deprived areas, covering 56% of the city's long-term unemployed and over a third of its territory. They are not the only response to neighbourhood decline and they all have different characteristics, reflecting diverse local circumstances and a different balance between top-down and bottom-up influences (Rodgers, 1995). They have all become more important over time and now represent a sizeable component of Glasgow's development capacity. They have all been established for over five years, which is long relative to many regeneration initiatives elsewhere, although not in terms of the scale of the problem. They currently employ 420 people in total (not counting trainees and those on temporary work placements) with a combined annual budget of £17m, derived from various local, regional, national and European sources. Their areas vary in size and character from large parts of the inner city with an economic base, vacant land for development and varied housing types (such as the East End and Glasgow North), to smaller residential estates with tighter boundaries and little undeveloped land (such as Castlemilk and Drumchapel).

The LDCs perform a range of functions according to local circumstances, their skills and ambitions, credibility and track record, and the support available from major funders. Figures 3.2 and 3.3 summarise their main activities. Those in areas with a poor physical infrastructure, where there are few other local agencies concerned with economic and employment matters, and that have been in existence for some time tend naturally to be larger and engaged in more diverse activities (such as Govan, the East End and Glasgow North). Conversely, those in areas with existing institutions, or in mainly residential areas, tend to be smaller and more focused (such as Pollok, Gorbals and Castlemilk). The activities of the LDCs fall into three broad categories: providing customised services, championing their areas, and pursuing development opportunities.

Providing Customised Services

An important reason for setting them up was to fill gaps in the jigsaw of local service provision or to make city-wide and national programmes better suited to local needs. Customised services are adapted to the requirements of individuals, firms or areas. The LDCs can treat problems in a more holistic way and deliver more integrated services because they are smaller and less affected by departmental boundaries and bureaucratic procedures. Policy is determined closer to the point of service delivery, so should be more flexible and responsive to user requirements. For example, more intensive, person-centred and pro-active support is provided for the unemployed than available through standard government schemes. Similarly, new and existing businesses receive a more accessible, personalised and rounded development service than provided by city-wide organisations.

A good example of innovation in Glasgow is the 'intermediate labour market' (ILM). This offers the long-term unemployed temporary work experience at a reasonable wage, in the course of providing socially useful services, coupled with vocational training and personal development. The Wise Group pioneered the idea with large-scale home insulation and landscaping projects offering hundreds of work placements and an impressive record of getting people into jobs afterwards. Glasgow Works extended the concept by acting as a broker for the LDCs and other agencies to develop their own projects. Glasgow Works conducts central negotiations to secure the funds and then explores project ideas with the local

LDC area	Castle-milk	Drum-chapel	East End	Easter-house	Govan	Gorbals	G'gow North	Pollok	Total
Employment information/advice	**	**	**	**	**	**	**	**	
Counselling/personal development	**	**	**	**	**	**	**	**	
Flexible adult learning facilities	-	**	**	**	**	*	**	-	
Help to access college training courses through ESF	**	**	**	**	**	*	**	**	
Small grants to access jobs, interviews or training	**	*	*	*	*	*	**	**	
Jobclub/workshop	-	*	-	-	-	-	-	**	
Targeted pre-recruitment training	**	**	*	**	**	*	*	*	
Job placement/recruitment service	**	**	**	*	**	**	**	*	
Employment subsidies to firms	**	**	**	**	**	*	**	*	
Construction industry training/job scheme	**	**	**	**	-	-	**	-	
Intermediate labour market projects (no of places)	** (30)	*(16)	*(16)	**(36)	**(50)	*(14)	*(16)	-	178
Childcare: no of places or subsidies currently available	-	64	102	n.a.	100	30	-	-	296
Outreach services (number of other sites/bases)	3	-	8	5	8	-	4	1	
Estimated number of clients placed into jobs in 1997	450	416	583	350	650	280	500	397	3,626

Key: ** = a significant feature/service (a relative and subjective assessment made by the LDCs themselves)

　　　　* = a feature/service that is present but not very significant

　　　　- = feature/service is not provided directly (in some cases provided by other organisations in the area, sometimes with the LDC's support).

Note: there has been some attempt to standardise the definitions used for clients into employment and other activities and outcomes - see text.

Source: Author's survey of LDCs carried out in February 1998, based on an original idea by Rodgers, 1995.

Figure 3.2 Main Human Resource Development Activities of Glasgow's Local Development Companies

LDC area	Castle-milk	Drum-chapel	East End	Easter-house	Govan	Gorbals	G'gow North	Pollok	Total
Support for business start-ups	**	**	**	**	**	**	**	-	
Number of start-ups in 1997	20	26	39	20	120	40	38	-	303
Advice/information to local firms	*	**	**	**	**	**	**	-	
Training support to local firms	*	**	**	*	**	*	**	-	
Financial support to local firms	*	**	**	**	**	*	**	-	
No. of local firms actively assisted in 1997+	10	60	140	50	90	60	132	-	542
Business club or network	-	**	**	**	**	*	**	-	
Targeted help for inward investors	**	*	-	**	*	*	**	-	
Support for community enterprise	*	-	-	*	-	-	*	-	
Subsidiary trading ventures	*	*	*	**	**	-	-	-	
Land and property acquisition	*	*	-	**	**	-	-	-	
Infrastructure provision/improvement	-	*	**	*	**	*	*	-	
Environmental improvement	*	-	**	**	**	**	-	-	
Property development	**	*	-	**	**	-	**	-	
Security measures	-	-	**	**	**	*	*	-	
Community development	*	-	-	**	**	-	**	-	
Improving image of area	*	-	*	**	*	*	*	-	
Measures to improve health	-	-	*	-	-	-	-	-	

Key: ** = a significant feature/service (a relative and subjective assessment made by the LDCs themselves)
   * = a feature/service that is present but not very significant
   - = feature/service is not provided directly   + = includes key/target clients plus those on management and training programmes.
   Source: Author's survey of LDCs carried out in February 1998, based on an original idea by Rodgers, 1995.

Figure 3.3 Other Activities of Glasgow's Local Development Companies

organisations. This sophisticated enabling approach offers many lessons for other areas. It may be more robust than a centralised model in the long-term since it allows projects to be designed closer to local circumstances and 'market-tested'. They can either grow organically or be replicated elsewhere, if successful, or be closed down if not, without causing major damage. The diversity also provides scope for greater experimentation within a certain budget and time-frame than a single large programme. Figure 3.4 describes four illustrative projects.

True GRIT is a Gorbals Initiative Glasgow Works project providing a low price but quality market research facility for local community groups and other organisations whilst training long-term unemployed people in administrative skills and survey techniques. Various local groups, Scottish Homes, the City Council and private consultants have used the facility to undertake surveys of residents in order to improve their services.

Govan Teleworks was developed by the Govan Initiative with support from Glasgow Works. It provides call centre services for local SMEs to assist them with their marketing activities. The project has been successful at generation revenues from selling its services while coincidentally fulfilling its work experience and training objectives.

BOSCO is a Glasgow Works project developed by Greater Easterhouse Development Company. It provides 20 places to train long-term unemployed people to teach a variety of activities in the local sports centre in the hope that this will help them get jobs in this expanding sector. It also seeks to break down religious prejudices by drawing in people from different backgrounds.

Castlemilk Economic Development Agency has developed a Glasgow Works project called *Playsport*, in conjunction with Glasgow City Council, involving 16 people engaged in various aspects of fitness leadership and promotion, and play leadership. An earlier Fitness and Wellbeing project undertaken on a pilot basis with the Greater Glasgow Health Board was modified to form Playsport after the first year, following evaluation.

Figure 3.4 Intermediate Labour Market projects

Local Champions

Most LDCs have become more ambitious over time and acquired the role of local catalysts for change or champions for their areas. This partly reflects a recognition of limits to what can be achieved by providing supplementary services themselves, since these are unlikely to produce the impact on their areas of redirecting main public sector programmes or attracting major investment projects.   Increasing competition between areas for jobs and resources has also made it more important for them to represent and advocate local interests in wider arenas.   The notion of a local champion can mean raising the area's profile to attract investment; representing local needs in external decision-making forums; pressing for increased support and flexibility from mainstream policies; seeking to enhance or reorient city-wide programmes to better suit the local situation; challenging strategic land-use policies to secure more flexibility for local development, and persuading city employers to recruit local people.   Figure 3.5 describes two illustrative projects.

Development Agents

The LDCs also carry out physical development work directly.   Their contacts with local enterprises and knowledge of the local economy generate insights into what needs to be done and stimulates useful project ideas. Their focused remit and local awareness helps to identify vacant land, old buildings, businesses with growth potential and people with particular skills that could be converted into more productive assets with advice and support. They may be able to progress these relatively quickly given their flat decision-making structures and freedom from the constraints of borrowing money, generating income and financial clawbacks that affect most public authorities.   As small organisations with an outward orientation they may develop a culture of action and experimentation.   Simple internal communication and decision procedures mean they can tackle issues in a more comprehensive and integrated way than larger bodies.   Part of the motive for undertaking development is also to increase local ownership of assets, since many LDC areas are council housing estates.   Figures 3.6 and 3.7 describe four illustrative projects.

A large industrial site in **Drumchapel** had been vacant since 1980 when Goodyear closed its tyre factory. There were various retail proposals for the site during the 1980s, but it was always reserved by the regional planning authority for industrial use. Yet it made no investment in infrastructure (e.g. service roads) and landscaping to make it marketable. In the early 1990s the Drumchapel Community Organisations Council and Drumchapel Opportunities (DO) worked with private developers on a mixed-use proposal involving a major food superstore, retail warehouses, a pub, bingo hall, industrial units and some private housing. After considerable negotiation and representations by supporters, this obtained approval from the planning authority and the development went ahead. Local residents gained a share of the jobs created after receiving customised training and job preparation from DO. Other arguments for the development were that local residents would benefit from lower food prices and greater choice, access to the leisure facilities, and it would improve Drumchapel's image. One of DO's roles in the negotiations was to ensure that the economic interests of residents received proper attention.

In the mid-1990s the Scottish Office granted planning permission for a major out of town shopping centre, with some complementary industrial development and leisure activity, at Braehead, straddling the Glasgow/Renfrew council boundary, against the wishes of the local authorities. As part of the 'planning gain', Glasgow City Council persuaded the developer to build some 3,000 sq.m. of industrial units and hand over the completed development to a third party to benefit from the rental income. The site selected was a former housing estate in central Govan that had been cleared by the Council with the intention of converting it to industrial use. **Govan Initiative** persuaded the Council to let it become this third party. After a lengthy period of negotiation they also obtained the right to actually undertake the development themselves, arguing that they knew the local market better and could raise ERDF co-funding to increase the scale of the project. The Moorpark Industrial Estate was completed in April 1997, consisting of nine units ranging from 100 to 1,000 sq.m. All are now occupied with leases of 15-25 years. Govan Initiative believes that its intervention increased the scale of the development and secured a larger income stream for reinvestment in the area and use as match funding for other projects.

Figure 3.5 Facilitating development through advocacy

Strathclyde Regional Council helped *Greater Easterhouse Development Company* to acquire a redundant school building which they converted into 1,800 sq.m. of high specification office space in Easterhouse Town Centre, at a cost of £2.1m funded from ERDF, the GDA and the Regional Council. Space in the Westwood Business Centre was made available for new tenants in a range of sizes and on flexible terms with full supporting services, including reception facilities, conference room, administrative support, security, cleaning and catering.

East End Partnership acquired an old public wash-house from Glasgow City Council in 1995 and converted the attractive listed red sandstone building into its administrative headquarters as well as an advice centre and space for training and business support facilities. The cost of £0.53m was met by ERDF, local authority and GDA funds. East End Partnership has also been involved in derelict land renewal and the building and refurbishment of industrial premises, usually in partnership with the GDA.

Figure 3.6 Property development - office space and training facilities

The *Gorbals Initiative* was involved in upgrading a 1960s industrial estate (Dixon's Blazes) with £1m from the GDA, ERDF and local companies in order to safeguard the 2,000 existing jobs in the area and encourage new investment. The scheme involved renovating industrial premises, landscaping, creating parking space and improving security. As a result, firms are consolidating their activities on the estate and investing substantially more themselves. The Gorbals Initiative is also helping them to improve their performance through business development support. Empty units on the estate are being let to new tenants as confidence improves.

CEDA developed the Glenwood Business Park on the site of a former school in Castlemilk which was demolished. It offers a range of new industrial units and office space amounting to about 4,500 sq.m., which was built at a cost of £5.2m from ERDF, the Scottish Office and GDA. The development was phased over a four-year period to test demand for the accommodation from local businesses and those outside the area. CEDA also provides support for tenants of the premises in the form of business planning, access to finance, signposting to other services and help with staff recruitment. On the same site CEDA also developed an integrated training facility providing 1,200 sq.m. of training space. Part of this is occupied by a new local annex of Langside College.

Figure 3.7 Property development - industrial and business space

Institutional Arrangements

The LDCs are legally constituted as companies limited by guarantee with charitable status, which gives them discretion to undertake a wide range of activities, but no statutory authority. They are formally controlled by boards of directors whose function is to make effective use of the relative autonomy of the LDCs and to give direction to their activities. Board membership reflects the distinctive focus of each LDC's activities and its unique relationships with other local and city-wide organisations, which generally include the local authority, development agency and representatives of the local community and private sector. This gives each LDC a unique 'organic' partnership character in that it is constituted by and accountable to a range of different organisations and interests, which gel and develop in diverse ways.

Board representation by key funders offers a closer, more sophisticated relationship of dialogue, support and influence than the conventional arms-length contracting model of accountability. It is flexible, enables certain central powers and responsibilities to be 'let go' in order to help get things done, and allows for local projects to evolve over time as the policy environment changes. Most other board members are co-opted and reflect particular local and historical circumstances, such as representatives of local colleges jointly involved in providing training. Some board members are chosen for their special expertise or connections to build alliances with wider organisations to benefit the local area. All the LDCs include community representatives and elected councillors on the board to facilitate communication and accountability to local residents. Public meetings, briefings, training and outreach services are also used to promote community involvement.

Few would claim that the LDCs are a panacea for removing local concentrations of unemployment and poverty, although they clearly perform many useful functions. Decentralisation allows greater responsiveness to local circumstances and a more holistic approach to economic development than normal. Partnership working brings different interests together in a typically constructive and creative manner. A relatively flexible and durable institutional framework has been established in most localities, although it will always be vulnerable to the vagaries of funding, politics and personalities. According to internal performance monitoring systems and independent evaluations of their programmes and procedures, significant progress has been made and additional economic impacts have been achieved in a cost-effective manner. It is more difficult to measure their overall impact on neighbourhood conditions, partly because of the scale of the problems and the technical complexity of such analysis.

Dublin's Experience

Dublin's context is one of regional economic growth and in-migration of population, in complete contrast to Glasgow. Yet, certain peripheral estates and inner city neighbourhoods have failed to benefit from recent prosperity and suffer from acute long-term unemployment and multiple deprivation. As a result, 11 'local partnerships' were established in different parts of the city between 1991 and 1996, part of a national programme to develop area-based regeneration strategies in deprived urban and rural localities across the country. A combination of bottom-up and top-down forces shaped this programme, including long-term lobbying by the community and voluntary sectors, support from the various social partners, especially the trade unions, pressure from the European Commission, and positive endorsement by the government itself.

Dublin's local partnerships are similar to Glasgow's LDCs in drawing together a range of partners in a voluntary capacity to focus on local economic and social development. They are registered as independent companies and their board membership includes the local community, voluntary sector, business, trade unions, local authorities and national training, education, social welfare and economic development bodies. This gives them some influence over selected agencies of central government, in theory bending and integrating mainstream programmes at the local level. They also have the right to bid for funding to provide services not currently delivered by the statutory bodies, which they have pursued in the fields of human resource development and business development.

The single most important group of partnership activities are those intended to reduce unemployment, through training, work experience, employment guidance and counselling, job search, employer liaison and vacancy matching. Other activities include measures to tackle; educational disadvantage (through all kinds of school-based, pre-school and parental support); environmental decay (through physical renewal and landscaping); poverty (welfare resource centres and advisory services); self-employment (business training, advice, guidance and financial support); and, community development (supporting networks of local organisations concerned with issues such as housing, drugs and unemployment). The scale of individual activities is smaller and more experimental than in Glasgow, since there is less experience to build upon. There is also generally less organisational capacity on the ground. In addition, the local partnerships relate more closely to agencies of central government and are largely independent of local government, partly because Irish local authorities are much weaker than in Britain.

The partnerships are just as different from each other as Glasgow's LDCs, reflecting local circumstances and internal dynamics, and a lack of central imposition. Some focus on a 'service delivery' type function, providing a range of relatively standard economic and social services that may not previously have existed or been utilised in the area, partly through the statutory government agencies. Others have progressed into an 'agency' role, whereby they are more concerned with the design and delivery of new and additional measures. A third group have adopted more of a 'brokerage' role (which is preferred by the centre), where the emphasis is on supporting community-based organisations to build their own skills and institutional capacity by assisting through planning, lobbying, co-ordination, facilitation and other catalytic activities. They are concerned to avoid the partnership structure becoming an end in itself, distracted by financial and bureaucratic imperatives, and losing sight of the broader development objectives.

The local partnerships each receive relatively modest direct funding of roughly £0.5m per annum. Most of this is from the EU under Ireland's Objective 1 Plan for 1994-99, co-funded by the government. Many partnerships supplement this by applying for additional project-specific funds and resources secured by drawing in mainstream government services. The EU funds are channeled through an intermediary organisation ADM Ltd. (Area Development Management) that was created by the government to manage the process of setting up the partnerships and assisting their development (ADM, 1995). ADM works closely with key government departments to facilitate their support.

It is too soon to judge the impact and effectiveness of the partnerships, especially as monitoring and evaluation are somewhat undeveloped. Simple performance measures based on participation in training programmes, job placements and business start-ups suggest sizeable quantitative effects. Informal evidence of community capacity building, institutional networking and national policy changes suggest equally important qualitative achievements. Yet, their long-term position seems less stable than Glasgow's LDCs, since EU funding is likely to be curtailed after 1999 and their relationship with local authorities is a source of growing tension and uncertainty (Walsh, 1998). Many are concerned that the imminent reform of local government may lead to their incorporation and loss of independence.

Duisburg's Experience

Duisburg's context is one of continuing deindustrialisation, rising unemployment and growing social and racial tensions in some of the city's poorest neighbourhoods. Manufacturing jobs in the city collapsed from 135,000 in 1976 to 66,000 in 1996, while employment in private services increased by only 6,000. With some notable exceptions, regeneration programmes in Germany have traditionally had a strong physical orientation and been organised by the public sector with limited involvement of the local community or private sector. This is beginning to change as a result of initiatives from the state (regional) and local authorities, with some encouragement too from the European Commission. Emphasis is being given to more comprehensive and integrated approaches to neighbourhood improvement, with some tentative forms of decentralisation and partnership working. Duisburg is at the forefront of such activities in several respects. Public authorities here are seeking to make economic and social development more important aspects of area regeneration.

These efforts are most advanced in Marxloh and Bruckhausen, two neighbourhoods in the north of the city hit hard by the decline of the steel industry and experiencing cumulative problems of dereliction, low income, population turnover, selective outmigration, racial segregation and poor housing. Since the early 1990s, the authorities have been pursuing a broad-based programme including renewal of the physical infrastructure, housing and environmental improvement, temporary employment and training schemes, job counselling and placement services, business start-up and growth assistance, improved social and welfare services, and recreational and cultural activities to encourage social interaction. This is as broad an agenda as in Glasgow or Dublin, although many of the economic projects in particular are on a smaller scale, reflecting their novelty.

The local authority has been leading these initiatives, backed by substantial financial and political support from the state government and EU (under the URBAN programme). Inter-departmental offices of the authority have been established within the neighbourhoods in order to achieve a more accessible local presence and to provide new and more coherent services than possible with the traditional departmental hierarchy. In Marxloh a limited company has also been created to promote economic and property development, wholly owned and controlled by the local authority. Such initiatives are expected to act as catalysts for local change by actively exploring and implementing development projects of various kinds. They are also expected to strengthen the skills, cohesion and organisational capacity of the community to promote greater self-reliance and economic independence, and to ensure that local improvements are sustainable.

In most cases the extent of community involvement and influence in decision-making is limited. There is a fair amount of informal communication and co-operation, but no officially recognised mechanisms for articulating and representing community needs and aspirations. Business interests also tend to be weakly represented. The official rationale for this has been the need for urgent action in a critical situation, but there is growing pressure for a more balanced relationship to be established between public authorities, local citizens, the private sector and intermediary organisations in shaping area regeneration efforts.

It is premature to draw definite conclusions about Duisburg's regeneration schemes, particularly as monitoring and evaluation systems are less developed than in Glasgow or Dublin. Preliminary indications are that the quantitative impacts to date are probably less significant than the qualitative effects. The creation of temporary jobs for local residents, opening of new facilities and support for new businesses provide tangible evidence of progress and offer a boost to local morale. They have not begun to replace the jobs lost though industrial decline, but indicate that something can be done and should help to encourage further activity. The poor image of the target neighbourhoods is expected to improve and stronger community networks should promote social cohesion and population stability. Regeneration policy in Duisburg is evolving fairly quickly, partly in recognition that there is scope for improvement in partnership working, increased commitment to local development activity, and improved co-ordination with wider city and regional policies.

Conclusions and Lessons

It seems highly unlikely that there is single best model or approach to regeneration across European cities. It is important that policies and structures are tailored to economic, social and institutional conditions. Nevertheless, there are still some broad lessons to be learnt. In the debate about mainstream bending or local capacity building, some cities are better placed than others to target investment and services towards local concentrations of deprivation through existing public authorities. Lack of physical capacity and resources, or the presence of more widespread economic and social problems across the city, may make the establishment of independent local development agencies a more fruitful approach in several respects.

This point is also relevant to whether local regeneration strategies should be primarily geared towards redistributing jobs, training and other opportunities towards disadvantaged communities, or contributing to city-wide prosperity by transforming underused property and other business and human assets into more productive units, and in the process securing additional investment, jobs and incomes for the city. In general there appears to be insufficient flexibility within national policies to accommodate such diversity in local approaches.

In most cases some combination of local/city and national/supra-national policy measures are likely to be important to benefit from the advantages of both, albeit with a different balance of emphasis depending on the circumstances. Local initiatives generally permit problems to be tackled in a more rounded and responsive way. However, wider policies have greater resources and more powerful levers of influence available. They can also help to reduce some of the fragmentation of effort and uneven development that results from local institutions varying in their effectiveness. The recent statement from the UK that: "The Government is convinced that the 'bottom-up' approach to regeneration is the right one" (DETR, 1998, p.2) is surely too simple and understates the Government's own potential contribution. Europe may also have a useful role to play in providing supplementary funding, supporting relevant research and policy learning, disseminating good practice and flagging-up impediments to local regeneration.

The structure of a community-based development company operating under the umbrella of a multi-agency partnership board provides a flexible, accountable and potentially powerful arrangement, with proven success in Glasgow. It is no panacea on its own, however, and still needs active support from the various partners at all levels and effective reinforcement by complementary wider policies and programmes. Other, more detailed practical lessons for local regeneration in Britain and elsewhere in Europe are discussed elsewhere (Turok, 1999). They include issues relating to the devolution of powers and responsibilities; more flexible and reliable funding mechanisms and administrative systems; strategic programmes of action; closer alignment between local and city-wide development policies; mechanisms for greater community and private sector involvement; increased collaboration between local partnerships, and greater attention to aspects of organisational development. There is clearly a great deal more to be done at all levels to promote regeneration and reduce the inequalities within many European cities and regions.

Acknowledgements

This chapter draws on a study sponsored by the European Commission and Glasgow Development Agency. The full and summary reports are available from the author on request. Considerable thanks are due to many individuals involved in local regeneration in all three cities for their information and insights - unfortunately too numerous to mention all by name. The steering group of Janice Roach, David Coyne, Mick Rodgers, John Watson, Brian McAleenan, Laurie Russell and Anne Brooks gave particular advice and guidance. Sean O'Siochrú helped greatly with the material on Dublin and Ralf Zimmer-Hegmann and Sabine Weck with Duisburg. The usual disclaimers apply. A different version of this chapter is to be published in *Local Economy*.

References

Area Development Management (1995), *Integrated Local Development Handbook*, Dublin.
Department of the Environment, Transport and the Regions (1997), *Regeneration Programmes - The Way Forward*, Discussion Paper, London.
Department of the Environment, Transport and the Regions (1998), *Community-Based Regeneration Initiatives*: A Working Paper, London.
OECD (1996), *Ireland: Local Partnerships and Social Innovation*, Paris: OECD.
Rodgers, M. (1995), Area Based Economic Development, Glasgow City Council, *Working Paper*.
Turok, I. (1999), *Inclusive Cities: Building Local Capacity for Development, Summary Report and Policy Guidelines*, Glasgow Development Agency and European Commission. Available from the author.
Walsh, J. (1998), 'Local development and local government in the Republic of Ireland' *Local Economy*, 12, 4. pp.329-341.

4 Regional Convergence in Italy and Spain: a Comparative Perspective

MARINELLA TERRASI

Introduction

This paper adopts an unusual approach in the study of regional convergence. In the past the evolution of regional disparities has been mostly analyzed in a strictly national context. In recent years, which have seen a revival of regional disparities studies, great attention has been given to the international scale and particularly, the European scale (Barro and Sala-i-Martin, 1991; Papers in Regional Science, 1995; Armstrong and Vickerman, 1995). In contrast to both these approaches, we shall compare the regional convergence experience of two countries, Italy and Spain, during their post World War II development.

The reasons why we think that a national comparative perspective may be useful in studying regional convergence are varied. First of all, there are some analytical advantages. The recent literature has pointed out the inconclusive results of empirical studies on regional disparities. In theory, it is clear that two different groups of factors are at work: the neo-classical equilibrating mechanism of declining marginal product of capital, and the disequilibrating effects of increasing returns obtained through internal and external economies of scale. In reality, the final results are determined by some specific factors, which are tied to the particular history and characteristics of each country and to the particular phase of growth of the international economy. Therefore, by comparing the experiences of different countries we may construct a sort of laboratory experiment and succeed in separating the general factors of the theory from the particular institutional, cultural and economic characteristics that have been relevant in each country.

Some more reasons that justify a comparative perspective in studying regional convergence touch the political and normative sphere. Among the specific factors of regional convergence there is also the strength and the form of regional policy. Therefore, the comparison of different national cases may help identify the most successful regional policy strategies. Moreover, it is our opinion that regional problems present a strong national dimension even in times of growing international integration. For this reason, a better knowledge of the regional/national relationship may serve to improve the effectiveness of regional policy at the European level as well.

For all these reasons, we have decided to confront the numerous statistical difficulties that arise when comparing the regional economic evolution of different countries, due to the heterogeneity of the sources of data and the characteristics of the analytical tools at our disposal. Even though our results are affected by such limits in some measure, we think that the bases of data and the statistical methods used make the comparison significant.

Our choice of Italy and Spain as objects of the experiment is not accidental. A comparison of these countries has been frequently made in the development literature (Fuà, 1980; Mingione, 1995; Prados de la Escosura and Zamagni, 1992) because they share some important characteristics, not only their geographical position and cultural and historical background, but also the delay in the industrialization process with respect to other European countries. Recently Mingione (1995) has discussed the hypothesis that both countries along with Greece and Portugal constitute variants on a particular model of capitalist development, in which family enterprises and self-employment have assumed an unusually important role. For our purposes the common base of development characteristics and experience in Italy and Spain makes it possible to point out some other conditioning factors which are considered important for regional convergence: displacement of the development phases, spatial structure, integration into the European Economic Community and international markets, institutional characteristics and in particular the different degree of administrative and political territorial decentralization.

The paper is organized as follows. First, the aggregate performance of the two countries in the period 1951-1992 is briefly analyzed with the aim of differentiating both the levels and rhythms of their respective development processes. In our line of reasoning this is not a secondary aspect because we believe that the aggregate experience has conditioned the regional convergence process. Nevertheless, we will be extremely concise in discussing the main characteristics of Italian and Spanish postwar development and limit our attention to those aspects that we consider relevant for the evolution of regional disparities. Next the evolution of regional disparities in Italy and Spain in the period 1955-1993 is analyzed through the Theil index of concentration. It is shown that both countries present a period of strong regional convergence, followed by a phase of divergence, with the difference that the Spanish process presents a delay of about ten years and looks more equilibrated from a spatial point of view. A relationship between regional disparities and national development is subsequently postulated and successfully estimated. The disaggregation of the Theil index of concentration in between- and within-groups of regions components follows. For the Italian case we build on the rich tradition of studies about the spatial dimension of post-war development (Arcangeli, Borzaga and Goglio, 1980; Bagnasco, 1977; Crivellini and Pettenati, 1989; Fuà and Zacchia, 1978) in order to select the appropriate groups of regions. For the Spanish case, there also exists evidence that three regions, Pais Vasco, Catalunya and Madrid, have guided the process of development (Cuadrado Roura, 1982; Suarez-Villa, Cuadrado Roura, 1993). The results of the analysis confirm the relevance of the spatial dimension in both countries, but a dualistic structure emerges as much more significant in the Italian case. A second kind of disaggregation of the Theil index in labour productivity and employment rate shares is also attempted for the period 1971-1993, with the aim of understanding the role of two important factors of regional convergence that have been stressed by the recent literature (Dunford, 1996). Also in this respect the Spanish process of regional development results more equilibrated. The analysis of apparent productivity differentials in the most recent period (1979-1993) is further developed by assessing the role of sectoral structure and of sectoral productivity differentials in the regional convergence process. A different behaviour of Italian and Spanish within sector disparities is verified. Finally some general conclusions on the relevance of the Italian and Spanish regional experiences are considered.

The Aggregate Performance of Italy and Spain, 1951-1993

The principal difference between the Italian and the Spanish development process after the end of World War II is one of timing and continuity. Not only did Italy start to achieve a phase of intense growth sooner, the so called "economic miracle", but its growth also continued uninterruptedly during the whole 1951-1993 period, even though we must recognize a significant slowing down after 1975, in conjunction with the first oil shock and the end of the Bretton Woods international monetary system. This is well documented in Table 4.1 by the average growth rates of Italian per capita GDP, which were above 4% until 1975 and fell to 2.5 and 2.0 subsequently.

Table 4.1 Growth rates of real GDP per capita and of its components, Italy and Spain, 1951-1993

Rate of growth (%)[a]	1951-1960	1961-1975	1976-1985	1986-1993
		Italy		
GDP per capita	4.7	4.1	2.5	2.0
Total GDP	5.3	4.7	2.8	2.2
Population	0.6	0.6	0.3	0.2
		Spain		
GDP per capita	3.5	6.0	1.0	3.1
Total GDP	4.3	7.0	1.8	3.3
Population	0.8	1.0	0.8	0.2

Note: [a] Log-linear adjusted.
Sources: processed from ISTAT data; Prados de la Escosura data.

The behaviour of Spain in the same period is quite different. In this case it is appropriate to distinguish four different phases of development (Prados de la Escosura and Sanz, 1995), with the first two spanning the intense phase of growth of the Italian economy. For this reason we have chosen to present the data of both countries by adopting the temporal intervals that are most significant for Spain. The first phase goes from 1951 to 1960 and is marked by the international isolation of the country, where an autocratic regime was in power. As a result, Spain was unable to benefit from the expansionary phase that the other European countries, and particularly Italy, were experiencing after the end of World War II; its rate of growth of per capita GDP was only 3.5 compared to 4.7 in Italy, (Table 4.1). The second phase starts in 1960 after a substantial change in Spain's economic policy, which gradually opened the country up to international trade. With a delay of ten years compared to Italy, Spain underwent a period of intense growth, the so-called "golden age", and presented an annual growth of 6% per capita GDP, substantially higher than Italy's. This phase was abruptly interrupted in the mid-70s, which coincided both with the end of Franco's regime and the first oil shock. For a decade Spain was engaged on two fronts: the economic crisis and the high political instability that characterized its transition to democracy. The general performance was unsatisfactory, with a growth rate of real GDP per capita of 1%. Finally, in the last period considered, the rhythm of growth caught up again coinciding with the entry of Spain into the European Community and a new expansionary phase of the international economy. But this time the growth phase was very short. A new recession started in the early 1990s, which lowered the average growth rate of per capita GDP to 3.1% for the whole 1986-1993 period. Table 4.1 also shows the growth rates of the components of per capita GDP: total GDP and population. While the evolution of total GDP is very similar to that of per capita GDP, the rate of population growth shows a rapid convergence between Spain and Italy in the last period considered, when both countries present the same average value of 0.2%.

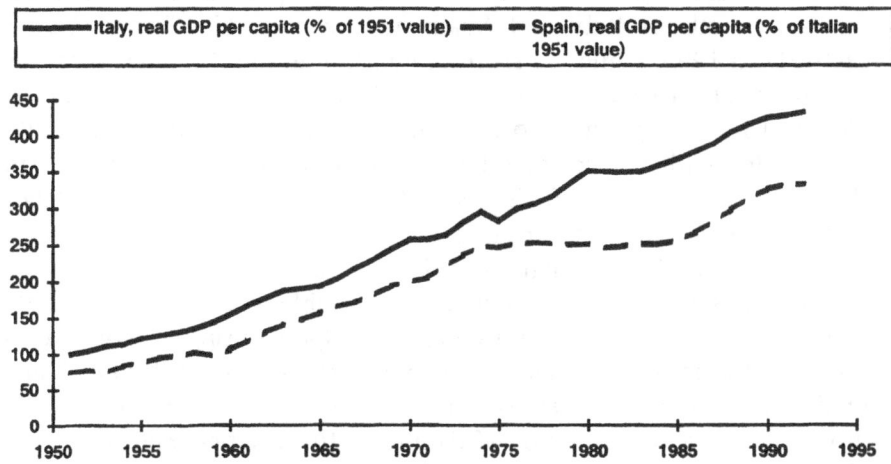

Source: processed from Summers and Heston data.

Figure 4.1 Italian and Spanish GDP per capita (1985 international prices)

We can evaluate the results of 40 years of development in Italy and Spain, by comparing the level of real GDP per capita at 1985 international prices, according to the latest estimates of Summers and Heston [1]. In Figure 4.1 we find the comparative performance of the two countries in the period 1951-1992. In 1951 Spanish GDP per capita was 75% of Italy's. By 1992 position of the two countries was unchanged, with Spanish GDP per capita at 77% of Italy's.   According to the same estimates, Spanish GDP per worker was 85% of Italy's in 1951 and still at the same level in 1992, while the labor force participation rate was substantially lower in Spain than in Italy both in 1951 (38.35% versus 43.64%), and in 1992 (36.35% versus 40.55%). This explains why the rate of Spanish GDP growth with respect to Italy's is lower in terms of population than of workers. We must also account for the much higher unemployment rate in Spain with respect to Italy (about 21% versus 11% in 1993 according to CEC, 1994), which makes the rate in terms of productivity of employed worker even higher.

While GDP per capita and per worker were growing, the productive structure of the two countries was changing substantially. In Table 4.2 we find the composition of output and employment in terms of four sectors: agriculture, industry, construction and services. We notice that in 1951 the weight of agriculture in output and employment was much higher in Spain than in Italy, but it became very similar in 1993, when it absorbed about 3.5% of output in both countries and 8.7% of employment in Italy versus 9.9% in Spain. The share of industry in output was substantially higher in Italy in 1951, but after that gradually diminished and in 1993 Spain's share was higher (25.6% versus 23.2%). Output in services grew continuously both in Italy and Spain and reached 67.4% and 62.6% respectively in 1993, while the share of construction in the same year was higher in Spain (8.3% versus 5.8%). The corresponding shares of employment make it possible to calculate the relative labour productivity of each sector with respect to total productivity, which are reported in Table 4.3[2]. At the end of the period considered the relative labor productivity results very similar in industry and services in both countries, but it is lower in agriculture and higher in construction in Spain.

On the basis of the data presented, we are able to conclude that Italy maintained an almost constant advantage over Spain during the forty years considered. Nevertheless Spain's accomplishments look substantial and surprising, given that they were obtained in a less open and more unstable environment. Also, the evolution of the productive sectors resulted in very similar patterns in both countries with an acceleration of Spain in the substitution of services for industry and agriculture in the most recent period.

Table 4.2   GDP and employment composition, Italy and Spain, 1951-1993

Share (%)	1951[a]	1960	1975	1985	1993
			Italy		
OUTPUT					
Agriculture	22.9	14.8	8.2	4.9	3.6
Industry	31.8	30.6	31.3	27.4	23.4
Construction	4.9	7.4	7.6	6.5	5.8
Services	40.5	47.3	52.8	61.2	67.2
EMPLOYMENT					
Agriculture	43.9	32.2	15.3	11.4	8.7
Industry	23.8	27.5	29.2	23.3	21.2
Construction	5.6	8.7	8.8	7.3	7.4
Services	26.7	31.6	46.7	58.0	62.6
			Spain		
OUTPUT					
Agriculture	30.7	23.6	10.1	6.0	3.5
Industry	23.5	30.9	30.3	29.4	25.6
Construction	3.4	3.9	8.1	6.3	8.3
Services	42.4	41.6	51.5	58.3	62.6
EMPLOYMENT					
Agriculture	50.0	42.3	23.8	18.2	9.9
Industry	20.2	21.8	27.3	24.5	22.4
Construction	5.5	6.7	9.7	7.3	9.4
Services	24.3	29.2	39.2	50.0	58.2

Note: [a] 1950 for Spain.
*Sources:*processed from ISTAT data; Prados de la Escosura and Sanz data.

Table 4.3   Evolution of relative sectoral productivity, Italy and Spain, 1951-1993

Sectoral productivity/ total productivity	1951[a]	1960	1975	1985	1993
			Italy		
Agriculture	0.52	0.46	0.54	0.43	0.41
Industry	1.34	1.11	1.07	1.18	1.09
Construction	0.87	0.85	0.86	0.89	0.78
Services	1.52	1.50	1.13	1.06	1.08
			Spain		
Agriculture	0.61	0.56	0.42	0.33	0.35
Industry	1.16	1.42	1.11	1.20	1.14
Construction	0.62	0.58	0.83	0.86	0.88
Services	1.74	1.42	1.31	1.17	1.07

Note: [a] 1950 for Spain.
Sources :processed from ISTAT data; Prados de la Escosura and Sanz data.

While this confirms the idea of a common path of development for the two Mediterranean countries, it leaves room for some differences and specificity in each of them, the rate of unemployment being one of the most impressive. Our attention must now turn to the regional convergence process, with the aim of comparing the spatial performance of the two countries during the same phases of development.

The Evolution of Regional Disparities

We have chosen to compare the regional convergence process of Italy and Spain by adopting one of the various concepts of convergence that have recently appeared in the literature (Barro and Sala-i-Martin, 1991; Sala-i-Martin, 1995; Quah, 1996): the concept of σ-convergence [3]. The reason we are leaving out the concept of β-convergence, on which the same literature has been concentrated, is that our main purpose is to link the regional convergence process to the national development process and for this purpose a measurement of the dispersion in regional distribution is more appropriate than the average speed by which poor regions reach rich regions in each country. In fact, in our approach we are more attuned to the line of regional research that was started in the 1960's by J.G.Williamson and subsequently revisited by many authors (Amos, 1988; Fan and Casetti, 1994; Fisch, 1984; Gilbert and Goodman, 1976; Krebs, 1982, Therkildsen, 1981; Williamson, 1965).

As for the particular measure of σ-convergence, we have chosen to use the Theil coefficient of concentration (Theil, 1967), which we consider particularly apt for comparing inequalities of different regional systems and, in addition, may be decomposed in some useful ways (Batty, 1974 and 1976; Walsh and O'Kelly, 1979). The index was calculated according to the following formula:

$$IC = \sum_i y_i \log(y_i \, / \, x_i) \qquad (1)$$

where IC is the Theil index, y_i is the share of nominal national product for region i, and x_i is the share of national population for the same region. The index was standardized by dividing (1) by its maximum value, which is $log(P)$, where P is the national population (Walsh and O'Kelly, 1979). In this way we obtain an index of regional inequality which ranges between zero, corresponding to perfect equality, and 1, corresponding to maximum inequality.

The values of the index were calculated for 20 administrative Italian regions and 17 Spanish autonomous regions [4]. Both levels correspond to NUTS 2 regions adopted by the Statistical Office of the European Community. The boundary lines may be found in Figures 4.2 and 4.3, where we have also reported two numbers for each region, corresponding to the values of their per capita product, measured in relation to the means of the respective countries, in 1955 and 1993 [5]. The data on Spanish regional product came from the series on the provincial distribution of national income. This has been generally published since 1955, every two years by the Bank of Bilbao. The last year data was made available, was 1993. For Italy the series starts from 1953, but it was necessary to use three different sources of data not perfectly homogeneous (Tagliacarne, various years; ISTAT, 1995; SVIMEZ, 1993) [6]. The path of the index in the years considered is presented in Figure 4.4 for both Italy and Spain. This figure also includes a fictitious line corresponding to the values of the Theil index

Figure 4.2 Boundary lines of 20 Italian administrative regions

of Spain delayed by nine years. In this way it is possible to compare regional inequalities in Italy and Spain at similar levels of development, represented by the values of real national per capita GDP estimated by Summers and Heston for Italy (Summers and Heston, 1991) [7].

Figure 4.3 Boundary lines of 17 Spanish autonomous regions

If we look, first, at the historical values of the index, the behaviour of Italy and Spain looks very similar if we abstract from the period of autarky that Spain experienced in the 1950s and during which Spanish regional disparities were diminishing. The opening up to international markets marks a first period of growing disparities in both countries, followed by a long phase of convergence, which is interrupted in the mid-1970s by the first oil shock. Subsequently, some oscillatory movements of the index may be recognized both in Italy and Spain, without any definite tendency to grow or decline. By turning to Spain's delayed series we are able to establish that Spain and Italy presented a very similar Theil index in 1955, and the two countries may be considered alligned in terms of real per-capita GDP. But after 1955 the Spanish delayed index is always beneath the Italian index. This is particularly evident in connection with the phase

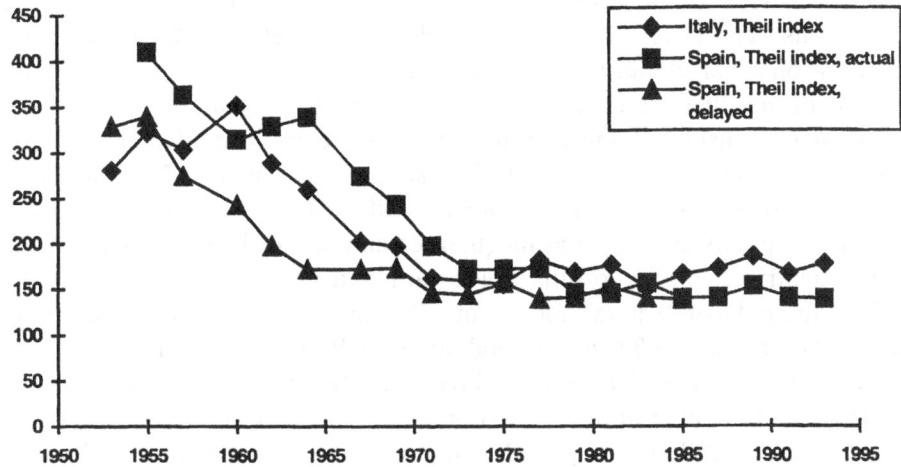

Source: processed from Banco de Bilbao data; ISTAT data; SVIMEZ data; Tagliacarne data.

Figure 4.4 Theil index (x100.000), total, Italy and Spain, 1955-1993

of intense growth that both countries experienced in the postwar era. It is mostly in this period that Italy accumulated a positive differential in the level of regional inequalities. This is subsequently reduced and almost eliminated in 1975. After this the Spanish index is always lower than the Italian one. Turning again to the historical series we verify a positive differential of the Italian index starting from 1983.

The general conclusion at this point of our comparative analysis is that the Italian and the Spanish regional convergence processes present a common general path in the 40 years considered and a better performance of Spain in two crucial periods of the respective national processes of development: a postwar era of intense growth and a phase of structural change that both countries underwent during the 1980s. This conclusion may be challenged by estimating a relationship between the Theil index and the level of development of each country. For this purpose, the following regression equation is proposed:

$$IC_t = a + bz_t + cz_t^2 + u_t \qquad (2)$$

where IC_t is the Theil coefficient in time t, z_t is the proxy chosen to represent the level of development, which has been made to coincide alternatively with real per capita GDP or time, and u_t is the error term.

The results of the estimation procedure may be found in Table 4.4. Equations (1) and (2) show that the relationship between the Theil index and national per capita GDP assumes a U-form both in Italy and Spain, which means that regional inequalities diminish less and less during the forty years considered and tend to grow at the end. The difference is that in Italy, the minimum point is reached around 1981, and in Spain in 1989. All the regression parameters satisfy and sustain the hypothesis of a relationship between regional inequalities and national development in the two countries. Similar results are obtained from equations (3) and (4), where the independent variable is time, but in this case the minimum point is reached around 1984 in Italy and again in 1989 in Spain. Equations (5) and (6) assume the GDP data estimated by Summers and Heston as the independent variable of the regression equation. In this case the results are almost unchanged for Italy with the minimum level of the Theil index falling around 1983, while for Spain the minimum is verified in the last year available, which is 1991.

Some interesting conclusions may be drawn from these results. First of all, we want to point out the fact that two countries such as Italy and Spain, which share some basic characteristics such as geographic position, strong regional differentiation and delay of the industrialization process with respect to other European countries (Fuà, 1980), have experienced a similar evolution in regional disparities during the forty years considered, notwithstanding the many differences in terms of political and institutional events. In our opinion this result supports the hypothesis of a strong relationship between the process of economic development and regional disparities. It is also remarkable that, within this common long term tendency, Spain achieved a more equilibrated regional distribution of growth, challenging us in the next part of our analysis to understand the reasons that might have determined this result.

The Spatial Components of Regional Disparities

A first contribution to the understanding of the different behaviour of regional disparities in Italy and Spain may be found in the geographical dimension that generally operates in the processes of economic development. In order to investigate this possibility we have decomposed

Table 4.4 Regression coefficients of equation (2) and Student's t (in parenthesis), Spain and Italy

Equation	Country	Dependent variable	Constant term	GDP	GDP^2	TIME	TIME^2	R^2 (Adjusted)	F
1	Spain	IC	591^c (18.79)	-18.04^c (-8.46)	0.18^c (5.56)			0.95 (0.95)	169^c
2	Italy	IC	498^c (13.43)	-40^c (-6.18)	12^c (4.81)			0.86 (0.84)	52^c
3	Spain	IC	957273^c (6.39)			-962^c (-6.34)	0,24^c (6.29)	0.96 (0.95)	175^c
4	Italy	IC	725056^c (3.66)			-730^c (-3.64)	0.18^c (3.61)	0.80 (0.77)	34^c
5	Spain	IC	619^c (14.06)	-97^c (-6.16)	4.87^c (3.70)			0.94 (0.93)	116^c
6	Italy	IC	528^c (10.83)	-69^c (-5.10)	3.30^c (5.84)			0.83 (0.81)	40^c

[c] Significant at 0,5%.

the Theil index of concentration into two additive parts: the between-group and the within-group inequalities, according to the following formulas:

$$IC = IC_{br} + IC_{wr} \tag{3}$$

$$IC_{br} = \sum_r Y_r \log(Y_r / X_r) \tag{4}$$

$$IC_{wr} = \sum_r Y_r \left[\sum_i (y_i / Y_r) \log(y_i / Y_r / x_i / X_r) \right] \tag{5}$$

where IC is total inequality, IC_{br} is inequality among groups of regions , IC_{wr} is inequality within groups of regions, y_i and x_i are the shares of national GDP and population for region i, and Y_r and X_r are the same shares for groups of regions.

The significant groups of regions have been chosen by taking advantage of the rich literature that exists both in Italy and in Spain about the spatial dimension of post-war development. In Italy (Arcangeli, Borzaga, Goglio, 1980; Bagnasco, 1977; Crivellini and Pettenati, 1989; Fuà and Zacchia, 1978) three different groups of regions can be usefully distinguished on the basis of the geographic position and the level of development in the 1950s: the North-Western triangle of Milan, Turin, Genoa (NW), which embraces the most developed regions in the 1950s (Val d'Aosta, Piemonte, Lombardia, Liguria); the regions of the North-East and Center (NEC), which are near the triangle and were characterized by an intermediate level of per capita GDP in 1955 (Trentino-Alto Adige, Veneto, Friuli-Venezia Giulia, Emilia-Romagna, Toscana, Umbria, Marche, Lazio); and the farthest and least developed regions of the South (S).

In Spain, too, we may distinguish some significant groups of regions, but in this case the delimitations proposed are more numerous because the behaviour of the autonomous regions during the forty years considered has been less homogeneous than in the Italian case[8]. We have chosen the delimitation generally used to analyze the period of intense growth in the 1960s and 1970s (Cuadrado Roura, 1982), because it presents many analogies with the one used for Italy. According to this, a first group of regions embraces Pais Vasco, Catalunya and the region of Madrid, which were the most developed in 1955 and acted as "fires of development" during the "golden age" of the 1960s and 1970s, attracting population and product. A second group includes the regions that at the beginning of the period presented a level of per capita product equal or above the national average and at the same time were positioned nearby the three "fires"; they are Asturias, Cantabria, Navarra, La Rioja, Aragon, Valencia and I.Baleares. The last group is formed by the peripheral and marginal regions of Galicia, Castilla-Leon, Castilla-La Mancha, Extremadura, Andalucia, Murcia and I. Canarias.

Table 4.5   Shares of national product and population for groups of regions, Italy and Spain, 1955-1993

Groups of regions	1955		1960		1975		1985		1993	
	GDP	POP.	GDP	POP.	GDP	POP.	GDP	POP.	GDP	POP.
Italy										
North-West	37.1	25.0	38.4	25.8	34.3	27.4	32.7	26.5	31.7	26.0
NEC	38.8	37.7	38.6	37.3	40.7	37.5	41.9	37.3	42.8	37.1
South	24.0	37.2	22.9	36.8	24.9	35.0	25.3	36.1	25.4	36.8
NC	75.9	62.7	77.0	63.2	75.0	64.9	74.7	63.8	74.5	63.1
Spain										
"Fires of devel"	38.4	23.8	37.8	25.7	43.7	33.7	42.0	33.8	41.9	33.8
Interm. regions	22.3	20.2	23.4	20.0	21.7	20.8	22.6	21.2	21.9	21.0
Underd. regions	39.2	55.8	38.7	54.2	34.5	45.4	35.3	44.9	36.1	45.0
Develop regions	60.7	44.1	61.2	45.7	65.4	54.5	64.6	55.0	63.8	54.9

Sources: processed from ISTAT data; SVIMEZ data; Tagliacarne data; Banco de Bilbao data.

In Table 4.5 the percentage shares of national nominal product and of population may be found for each group of regions and for some selected years. In 1955 the weight of the Italian "triangle" is very similar to that of the Spanish "fires": 37% versus 38% for product and 25% versus 24% for population. Twenty years later, at the end of the "golden age" of the Spanish economy, the same percentages were 34% versus 44% and 27% versus 34% respectively, showing that the polarization process was much stronger in the case of Spain. After 1975 the shares of population and product of this group of regions changed marginally in both countries.

The shares of the other two groups of regions are substantially different in the two countries, with the intermediate regions playing a more relevant role in Italy, where they absorbed 39% of product and 38% of population in 1955 versus 43% and 37% in 1993. In Spain the weight of the intermediate regions remains around 22% for product and 21% for population for the whole period. The loss of share of population presented by the marginal regions of Spain, which absorbed 56% in 1955 versus 45% in 1993, is rather impressive. Nothing of a comparable size happened in Italy, where the Southern regions absorbed 37% of population in 1955 and it was still 36.8% in 1993. The behaviour of the different groups of regions suggests that a simple bipartition might have gained significance at the end of the period. For this reason, in Table 4.5, we also show the shares of product and population obtained by aggregating the first two groups of regions, both in Italy and in Spain.

The results of the decomposition of the Theil index according to formulas (3)-(5) are presented in Figures 4.5 and 4.6 using the different groupings introduced in Table 4.5. By looking first at the results obtained when working with three groups of regions, we are able to conclude that in both countries the between-group component was particularly significant in 1955, absorbing about 93% of total inequality. The evolution of this component is similar to that of total inequality: it diminished both in Italy and in Spain, but at a higher rate in Spain and by 1993 the share of total inequality that it absorbed had decreased to 78% in Spain, while it remained constant at 93% in Italy.

The comparison with the results obtained when using the decomposition in two groups of regions is particularly interesting. In this case, while at the beginning of the period the between-group component absorbed a higher share in Spain (78% versus 70%), in 1993 it absorbed 91.6% of total inequality in Italy versus 68% in Spain. The corresponding line in Figure 4.5 shows that the division into two groups of regions had gradually gained relevance in Italy and at the end of the period it had perfectly substituted the division in the three groups.

The results obtained in this phase of our analysis confirm that in both Italy and Spain the regional convergence process presents a strong territorial dimension, which worked through the emergence of some poles of development, the gradual diffusion of growth to their neighbouring

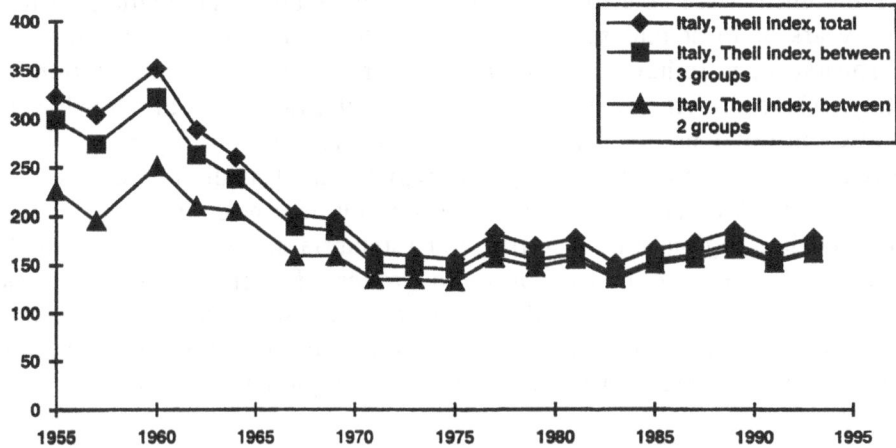

Source: processed from ISTAT data; SVIMEZ data; Tagliacarne data.

Figure 4.5 Theil index (x100.000), total, between 3 groups of regions, between 2 groups of regions, Italy, 1955-1993

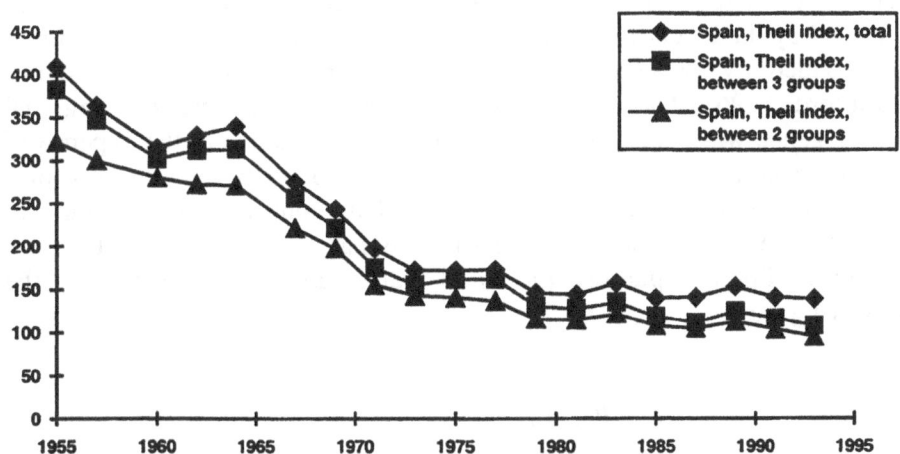

Source: processed from Banco de Bilbao data.

Figure 4.6 Theil index (x100.000), total, between 3 groups of regions, between 2 groups of regions, Spain, 1955-1993

regions and the difficulty of the peripheral regions to join the process of development. Nevertheless, some relevant differences emerged that may help to explain the more equilibrated behaviour of Spain noticed in section

2. First of all, the redistribution of population played a very limited role in Italy, contrary to what happened in Spain, where the "fires" of development were able to absorb a consistent share of population during the years of most intense growth [9]. In addition, the process of regional diffusion seems to have followed a more definite geographical direction in Italy, where growth gradually spread from North-west to the North-east and Centre regions, leaving the South completely cut out. By contrast, in Spain the position of some peripheral regions like Castilla-Leon, Castilla-La Mancha, Murcia and Canarias substantially improved, and some of the intermediate regions like Asturias and Cantabria substantially worsened, making the division in two or three groups of regions gradually less significant. On the whole, the Spanish economic system comes out geographically more mobile than the Italian one and this could in part explain its better performance in terms of total regional convergence.

Productivity and Employment Rate Components

We turn now to a different dimension of the regional convergence process: the contribution of productivity and employment rates to the equilibrating tendencies of regional per capita GDP. The role of productivity has been stressed by the recent literature; not only is the neoclassical model centred on productivity convergence (Wolf, 1994), but there is also some evidence (Dunford, 1996) of a trade-off between the two components of regional convergence with gains in relative productivity obtained at the expense of employment and *vice versa*. Our aim, then, is to verify the existence of a similar trade-off in Italy and in Spain and its contribution in helping to identify the causes of the different regional performance of the two countries. For this purpose the Theil index may be decomposed in the following way:

$$IC = IC_{pr} + IC_{emp} \tag{6}$$

$$IC_{pr} = \sum_i y_i \log(y_i / n_i) \tag{7}$$

$$IC_{emp} = \sum_i y_i \log(n_i / x_i) \tag{8}$$

where *IC* is total inequality, IC_{pr} is the share of total inequality due to regional differences in labor productivity, IC_{emp} is the share due to regional differences in employment rate (=ratio of employed to resident population), y_i, n_i and x_i are the shares of national product, employment and population of region *i*. As in the preceding sections the index has been standardized by dividing it by $log(P)$, where *P* is national population.

We applied formulas (7) and (8) to the 20 Italian administrative regions and the 17 Spanish autonomous regions considered in the preceding sections, obtaining the results reported in Figures 4.7 and 4.8. Unfortunately, we were forced to limit the period considered to the years 1971-1993 due to the substantial changes which occurred in the employment series of the Italian Regional Accounts starting from 1980; we made use of the data corrected by SVIMEZ (1989) which are available for the period 1970-1979. For Spain we used the number of jobs furnished by the Bank of Bilbao series rather than the employed population, because these data sets appear more suitable for measuring the labor input (Raymond and Garcia, 1994) and more homogeneous with the Italian data.

The Italian results show that productivity disparities remain almost unchanged after 1975. As a result, the behaviour of total disparities is essentially determined by the employment rate, which in 1993 absorbs 80% of the total Theil index. This confirms the idea that a limit exists under which the convergence of regional productivity cannot fall due to insurmountable differences in production conditions and structures (Raymond and Garcia, 1994). This limit, which resembles the steady state of neo-classical models, seems to have been reached in the Italian case by the mid-1970s. For this reason we cannot verify the existence of a trade-off, but rather the exclusive role of the employment rate in determining the regional convergence/divergence process.

The behaviour of Spain in the same period is quite different, as it is possible to check in Figure 4.8. In this case we see regional productivity disparities diminishing until 1991, and from 1975 the rate of employment tends to diverge. The trade-off appears verified and may be explained with the lower level of development in Spain compared to Italy for the same period. But by the end of the period, Spanish disparities in productivity have become very similar to the Italian ones (38 versus 36), while those of employment remain substantially lower in Spain (101 versus 142). We are able to conclude that one cause of the better regional performance of Spain with respect to Italy may be found in its greater capacity to control the regional disparities in the rate of employment. In this way our analysis confirms what Dunford has already demonstrated at the European level that "the rate of employment is often as important as the rate of productivity in determining a region's economic wealth" (1996, p.340).

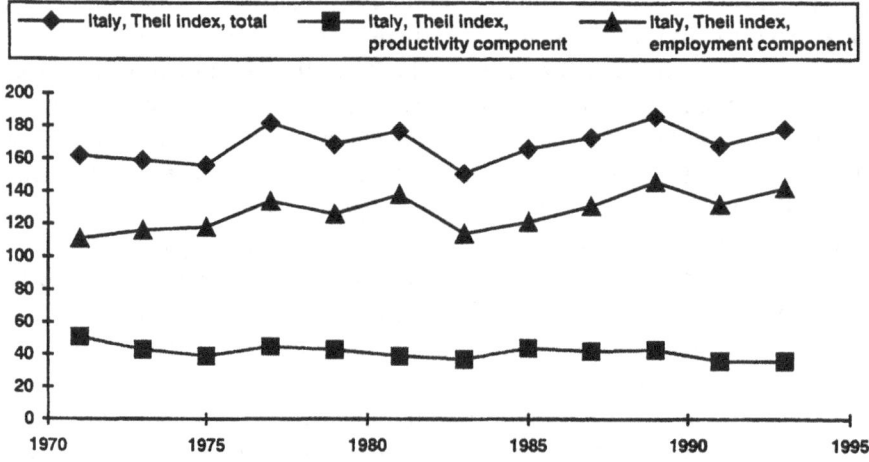

Source: processed from ISTAT data; SVIMEZ data.
Figure 4.7   Theil index (x100.000), total, productivity component and employment component, Italy, 1971-1993

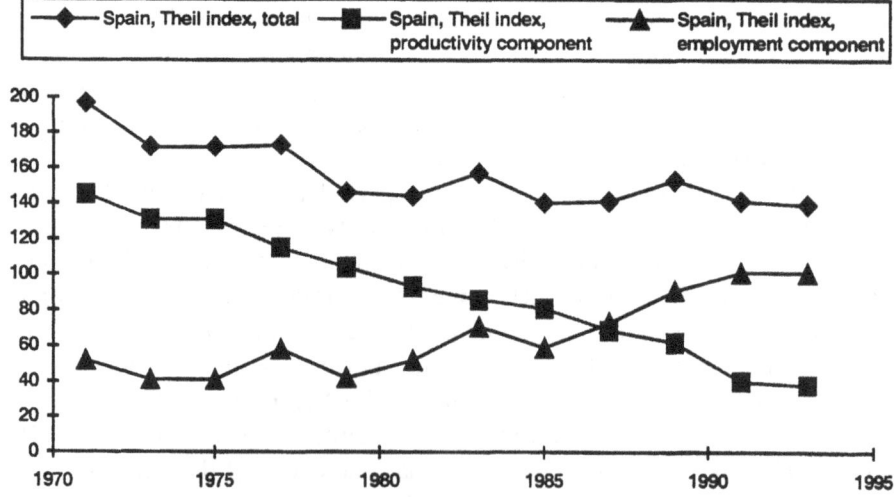

Source: processed from Banco de Bilbao data.
Figure 4.8   Theil index (x100.000), total, productivity component and employment component, Spain, 1971-1993

The Role of Sectoral Structure

Some additional understanding of the role of productivity in the convergence process of per capita GDP may be gained by developing a sectoral disaggregation. In this way we are able to distinguish between two different sources of productivity convergence: the reduction of productivity disparities within each sector and the equalization of the sectoral structure among regions. The disaggregation is particularly important when agriculture absorbs a relevant share of regional GDP, due to the much lower productivity of this sector with respect to industry and services. In Table 4.3 we can verify that this is indeed the case for the Italian and the Spanish economy at the aggregate level.

The attempt to isolate the role of the sectoral structure and sectoral productivity differences is not as straightforward as it was in the preceding decompositions. We have chosen, for our purposes, to readjust the procedure suggested by De la Fuente (1996) in order to get a synthetic indicator of the evolution of regional structure in each region. We have constructed a *virtual economy* by applying the regional sectoral productivities of the initial year to the observed sectoral employment data of the subsequent years. The result is a time series of total *virtual regional product*, on which we can calculate the Theil index. This should capture the evolution of sectoral structure differences among regions.

The outlined procedure was applied to the Italian and the Spanish data of regional value added at factor prices in four sectors of production: agriculture, industry, construction and services. The sources of data are the same as those used in the preceding sections. Again, we have some problems with the Italian data on sectoral employment, which at the sectoral level adopted are available on a comparative basis only from 1980. For this reason the time interval of our analysis is further reduced and goes from 1979 to 1993 for Spain and from 1980 to 1993 for Italy. The results are reported in Figure 4.9 [10]. They show that at the end of the period Spain has higher level of disparities in sectoral structure compared with Italy and, that in both countries, a tendency to render the sectoral structure homogeneous is at work.

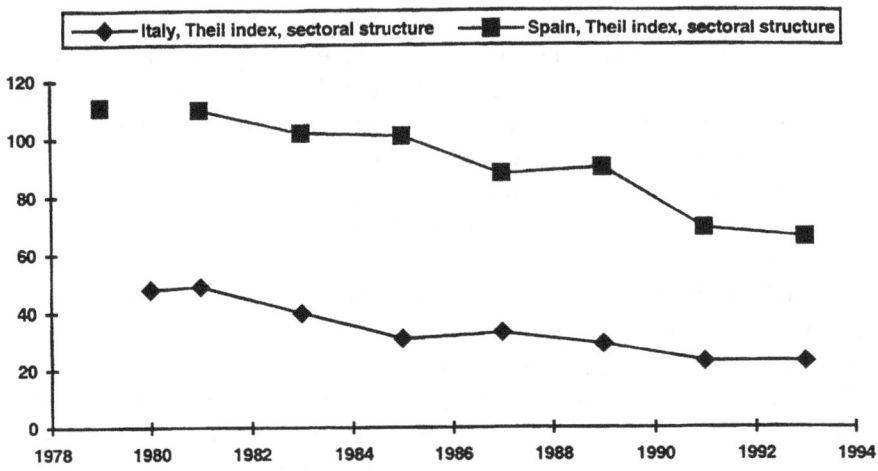

Source: processed from ISTAT data; Banco de Bilbao data.

Figure 4.9   Theil index (x100.000), sectoral structure, Italy and Spain, 1979-1993

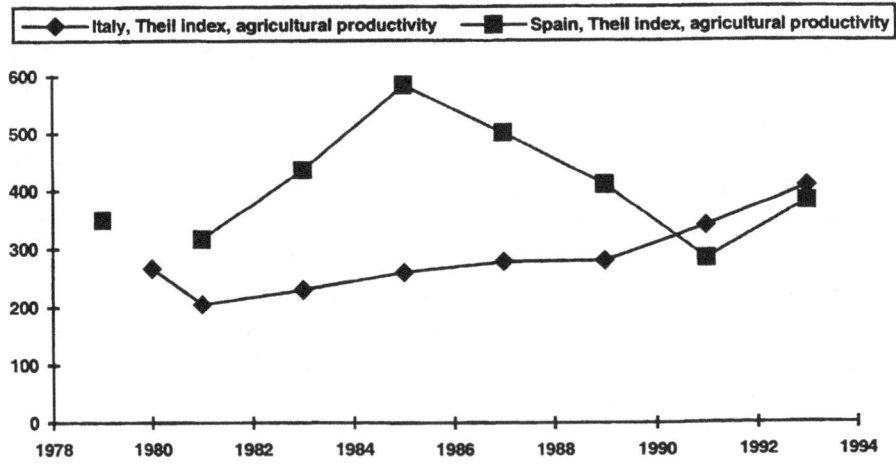

Source: processed from ISTAT data; Banco de Bilbao data.

Figure 4.10   Theil index (x100.000), agricultural productivity, Italy and Spain, 1979-1993

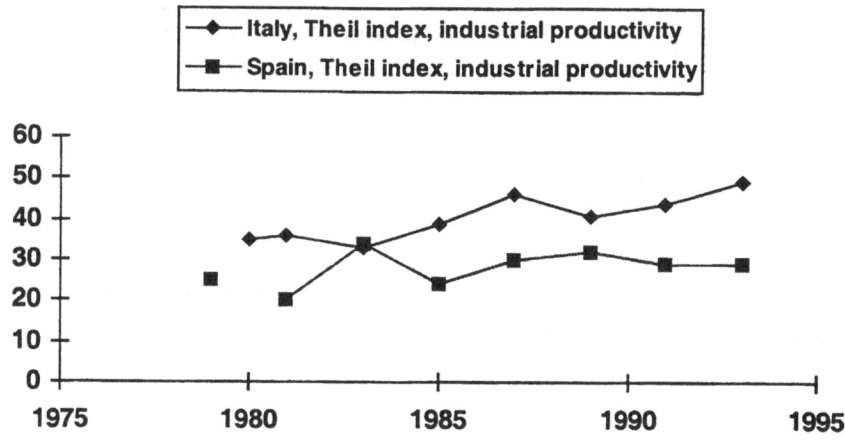

Source: processed from ISTAT data; Banco de Bilbao data.

Figure 4.11   Theil index (x100.000), industrial productivity, Italy and Spain, 1979-1993

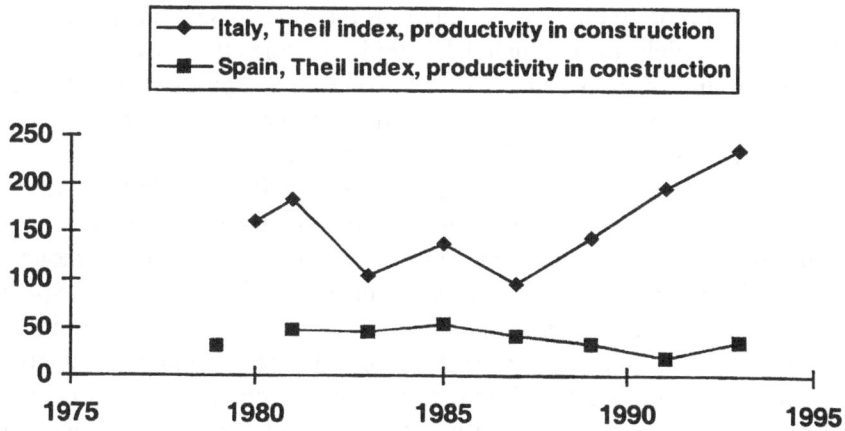

Source: processed from ISTAT data; Banco de Bilbao data.

Figure 4.12   Theil index (x100.000), productivity in construction, Italy and Spain, 1979-1993

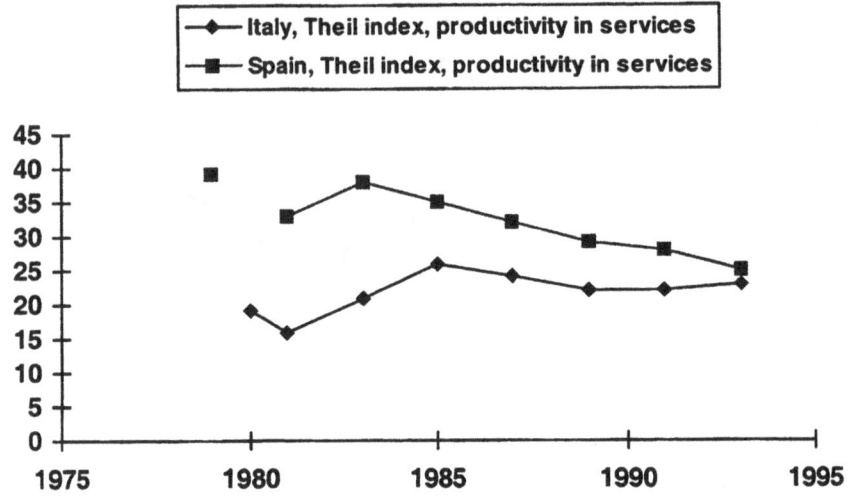

Source: processed from ISTAT data; Banco de Bilbao data.

Figure 4.13   Theil index (x100.000), productivity in services, Italy and Spain, 1979-1993

Quite different results are obtained on the side of sectoral productivity differentials, which are evaluated by applying formula (7) to the data of value added and of employment in each sector [11]. Figures 4.10-4.13 show that in this case at the end of the period considered the Spanish index is very similar to the Italian one for agriculture and services, while it is lower in industry and construction. Moreover, while in Italy the index had been growing in each sector, in Spain it had remained almost constant in the agriculture, industry and construction sectors and had decreased among the services. Therefore, the different performance of the Italian and Spanish total productivity divergence in the period 1980-1993 may be in part explained by the worse performance of Italian regional productivity differentials within each of the four sectors considered. This result confirms once more the less equilibrated regional evolution of the Italian economy.

Conclusion

The analysis of regional convergence that we have developed in the preceding sections for Italy and Spain has been characterized by its descriptive nature. Nevertheless, by using a simple regression analysis and different decompositions of the index adopted to measure regional disparities, we have been able to establish a relationship between regional convergence and some characteristics of each country's development process.

First of all, we have advanced the hypothesis that Spain's delay in opening up to the international markets and joining the European Community favoured the more equilibrated development path that we have verified with respect to Italy in the first part of the period considered. At the same time, by taking the long-term perspective, 40 years of postwar development, Spain's loss in terms of national growth does not look so bad, given that according to the last international estimates of Summers and Heston (1991), its per capita GDP had improved slightly compared to Italy by the end of the period. This suggests that the loss of efficiency in Spain was mostly restricted to the short term, and has been subsequently compensated by greater participation of all Spanish regions in national growth.

The geographical dimension of development has emerged as another important factor in the process of regional convergence of the two countries. In the early 1950s both Italy and Spain promoted a group of more advanced regions favoured by the geographical position in relation to the European border. In addition, Spain promoted an advanced region in the middle of the country: the region of Madrid. These two groups of regions have acted as centres of attraction for population and product in both countries. But while in Italy the diffusion of development has followed a quite definite geographical pattern moving towards the North-eastern and the Central regions and leaving the South cut off, in Spain it has presented some different characteristics. First, the redistribution of population has played a more important role and has substantially altered the weight of the less developed regions. Secondly, the behaviour of the intermediate regions has been more heterogeneous, with some of them losing and others gaining economic weight. On the whole, the behaviour of Spain looks more discontinuous than the Italian case. In our view this has made it possible for Spain to avoid the division of the country into two separate parts, which has been so harmful in the Italian case, and thereby realize a higher speed in regional convergence.

The last part of our comparative analysis of the Spanish and the Italian regional convergence processes is concerned with the role played by productivity and employment rate disparities. In this case we have profited from the conclusions reached in previous studies on European and Spanish regional disparities. According to these works there exists a trade-off between productivity and employment rate (Dunford, 1996) and a limit under which disparities in productivity cannot fall (Garcia and Milà, 1994). In addition we have been able to connect the worst Italian performance in the most recent years to the behaviour of the employment rate and to that of the productivity disparities within each of the four sectors in which GDP has been disaggregated.

This conclusion leads us to question what might have caused the different regional performances of Italy and Spain in the last part of the period considered. Neither the delay in the opening up to international markets nor the migratory movements of the population, which were indicated as important factors of differentiation for the initial phase of postwar development, appears to be workable explanations.

We advance the hypothesis that a new important factor of differentiation began to operate after 1978 with the enforcement of the new Spanish Constitution, which provided the 17 Spanish Regional Communities with a high degree of political and administrative autonomy. This in contrast to what has been maintained in Italy, where a strong centralist role for the state has been advocated in order to avoid an even greater marginalization of the Southern regions. In Spain the process of decentralization has not exacerbated growing regional disparities. It seems rather to have brought some beneficial effects for the spatial evolution of the Spanish economy.

In the mid 1980s a fortunate conjunction of events occured for Spain: the integration into the European Community and the full operation of the new political-territorial system. On the one hand, the need to compete with the strong nations of the European Community led the country to follow a general strategy based on efficiency and strengthening the external links of more advanced Spanish regions, yet at the same time the quasi-federalistic organization of the state introduced some interregional equity constraints and favoured the link of the less developed regions with the most dynamic areas (Suarez-Villa and Cuadrado-Roura, 1993). The role of infrastructure investment and of public capital in the spatial convergence process in Spain after 1980 has been verified in different studies (Mas, Maudos, Perez and Uriel, 1995; De la Fuente and Vives, 1995). Furthermore, Spain's entrance into the European Community saw the reinforcement and transformation of European regional policy along lines that were particularly attuned to the new decentralized organization of the Spanish state.

The same process of integration into the European Community had happened in Italy 30 years earlier in very different conditions. This was due to the highly centralized structure of the Italian state and to the different strategies of regional policy that prevailed at the time (Sutherland, 1986).

Does our comparative analysis of the Italian and Spanish experience in regional convergence provide some useful indications for future actions and strategies to be implemented? Our principal message seems to be not to undervaluate the regional dimension in the process of development. The Italian case shows that a country may be very successful nationally without being able to solve its regional problems, while the Spanish case suggests that in the long run it may pay to sacrifice some efficiency goals to realize a more equilibrated and cohesive pattern of growth.

Notes

1 The data were received on line via Internet. For a general description of the methodology see Summers and Heston, 1984 and 1991.
2 Relative labor productivity was obtained by dividing sectoral output shares by the corresponding employment shares.
3 For σ-convergence the dispersion of regional per capita GDP levels must decrease over time, while for β-convergence poor regions must grow faster than rich ones. Two concepts of β-convergence are introduced: absolute and conditional β-convergence. In the first case all regions approach the same steady state per capita GDP, while in the second case different regions have different steady states determined by their own structural characteristics, such as propensity to save, rate of growth of population, levels of technology, institutions and so on, and therefore one has to control in some way the steady states before testing the hypothesis of β-convergence.
4 We have not considered the Spanish region of Ceuta-y-Melilla because the corresponding data were not available for the whole period considered.
5 In 1955 we have only the aggregate product for the two Italian regions of Abruzzo and Molise. For this reason we find only one value (66) for both regions in Figure 4.2.
6 For Spain the data used are those of Gross Value Added (GVA) at factor prices; for Italy it was possible to use the same type of data for the period 1953-1969, but after 1969 it was necessary to use the data of Gross Domestic Product (GDP) at market prices, which make it possible to reconstruct a homogeneous series according to the new criteria adopted by the Italian Regional Accounts since 1980.
7 According to Summers and Heston estimates, Spain reached in 1960 the level of real per capita GDP that Italy reached in 1951.
8 Some delimitations derive from the European Community's regional policy objectives (Suarez-Villa and Cuadrado Roura, 1993); others distinguish the geographically best situated regions of the northeast quadrant of Spain and Madrid (Mas, Maudos, Pérez, Uriel, 1995) from the remaining regions; still others introduce the idea of an homogeneous Spanish Mediterranean Axis (Artis, Lopez-Bazo, Suriñach, 1997).
9 Raymond Bara and Garcia Greciano have analyzed the contribution of interregional migrations to the Spanish regional convergence process (Raymond Bara and Garcia Greciano, 1996).
10 In this case the standardization of the Theil index has been obtained by dividing it by $log(N)$, where N is the total number of labor units in the nation.

11 The Theil index was standardized by dividing it by the *log* of the total number of labor units in each sector in the nation.

References

Amos, O.M. (1988), Unbalanced regional growth and regional income inequality in the latter stages of development, *Reg. Sci. Urban Econ.*, vol. 18, pp. 549-566.

Arcangeli, F., Borzaga, G. and Goglio, S. (1980), Patterns of Peripheral Development in Italian Regions, 1964-1977, *Papers Reg. Sci. Ass.*, vol. 44, pp. 19-34.

Armstrong, H.W. and Vickerman, R.W. (eds) (1995) *Convergence and Divergence among European Regions*, London: Pion.

Artism, M., Lopez-Bazo, E. and Suriñach, J. (1997), Is There An Homogeneous Spanish Mediterranean Axis?, *Papers Reg. Sci.*, vol. 76, pp. 91-113.

Bagnasco, A. (1977), *Tre Italie. La problematica territoriale dello sviluppo italiano*, Bologna: Il Mulino.

Banco de Bilbao (or Banco de Bilbao-Vizcaya) (various years), *Renta Nacional de España y su distribución provincial*, Bilbao: Banco de Bilbao.

Barro, R. and Sala-I-Martin, X.X. (1991), Convergence across states and regions, *Brookings Papers*, vol. 1, pp. 107-82.

Batty, M. (1974), Spatial Entropy, *Geographical Analysis* vol. 6, pp. 1-32.

Batty, M. (1976), Entropy in Spatial Aggregation, *Geographical Analysis*, vol. 8, pp. 1-21.

CEC (1994), *Competitiveness and Cohesion: Trends in the Regions*, Fifth Periodic Report on the Social and Economic Situation and Development of the Regions in the Community, Luxemburg: Office for Official Publications of the European Communities.

Crivellini, M. and Pettenati, P. (1989), Modelli locali di sviluppo, in Becattini, G. (ed), *Modelli locali di sviluppo*, Bologna: Il Mulino.

Cuadrado Roura, J.R. (1982), Regional Economic Disparities: an Approach and some Reflections on the Spanish Case, *Papers Reg. Sci. Ass.*, vol. 49, pp. 113-132.

De la Fuente, A. (1996), On the sources of convergence: a close look at the Spanish regions, *Discussion Paper Series*, No. 1543, London: CEPR.

De la Fuente, A. and Vives, X. (1995), Infrastructure and education as instruments of regional policy: evidence from Spain, *Economic Policy*, pp. 13-51.

Dunford, M. (1996), Disparities in Employment, Productivity and Output in the EU: The Roles of Labour Market Governance and Welfare Regimes, *Reg. Stud.*, vol. 30, pp. 339-357.

Fan, C.C. and Casetti, E. (1994), The spatial and temporal dynamics of US regional income inequality, 1950-1989, *Ann. Reg. Sci.*, vol. 28, pp. 177-196.

Fisch, O. (1984), Regional Income Inequality and Economic Development, *Reg. Sci. and Urban Econ.*, vol. 4, pp. 89-111.

Fuà, G. (1980), *Problemi dello sviluppo tardivo in Europa*, Bologna: Il Mulino.

Fuà, G. and Zacchia C. (eds) (1983), *Industrializzazione senza fratture*, Bologna: Il Mulino.

Gilbert, A.G. and Goodman, D.E. (1976) Regional Income Disparities and Economic Development: A Critique, in Gilbert, A.G. (ed), *Development Planning and Spatial Structure*, pp. 113-141, New York: Wiley.

ISTAT (1987), *Annuario di contabilità nazionale*, vol. 15, Roma: ISTAT.

ISTAT (1992), Conti economici nazionali, *Collana d'informazione*, vol. 11, Roma: ISTAT.

ISTAT (1995), Conti economici regionali 1980-1992, *Collana d'informazione*, vol. 14, Roma: ISTAT.

Krebs, G. (1982), Regional inequalities during the Process of National Economic Development: A Critical Approach, *Geoforum*, vol. 1, pp. 71-81.

Mas, M., Maudos, J., Perez, F. and Uriel, E. (1995), Public capital and convergence in the Spanish regions, *Entrepreneurship & Regional Development*, vol. 7, pp. 309-327.

Mingione, E. (1995), Labour market segmentation and informal work in southern Europe, *European Urb. and Reg. Studies*, vol. 2, pp. 121-143.

Papers in Regional Science (1995), *Theme Issue: Regional Disparity*, vol. 74, pp. 87-204.

Prados de la Escosura, L. (1995), *Spain's gross Domestic product, 1850-1993*, Direción General de Planificación Working Paper, Madrid: Ministerio de Economia y Hacienda.

Prados de la Escosura, L. and Sanz, J.C. (1995) Growth and Macroeconomic Performance in Spain, 1939-1993, *Discussion Paper Series*, No. 1104, London: CEPR.

Prados de la Escosura, L. and Zamagni, V. (1992), *El desarolo economico en la Europa del sur. España e Italia en perspectiva histórica*, Madrd: Alianz.

Quah, D. (1996), Empirics for economic growth and convergence, *European Econ. Rev.*, vol. 40, pp. 1353-1375.

Raymond, J.L. and García, B. (1994), Las disparidades en el PIB per capita entre comunidades autonomas y la hipotesis de convergencia, *Papeles de Economia española*, vol. 59, pp. 37-58.

Raymond Bara, J.L. and García Greciano, B. (1996), Distribución regional de la renta y movimientos migratorios, *Papeles de Economia Española*, vol. 67, pp. 185-201.

Sala-i-Martin, X.X. (1995), The Classical Approach to Convergence Analysis, *Discussion Papers Series*, No. 1254, London: CEPR.

Suarez-Villa, L. and Cuadrado Roura, J.R. (1993), Thirty Years of Spanish Regional Change: Interregional Dynamics and Sectoral Transformation, *Int. Reg.Sci. Rev.*, vol. 15, pp. 121-156.

Summers, R. and Heston, A., (1984), Improved international comparisons of real product and its composition: 1950-1980, *Rev. Income and Wealth*, vol. 30, pp. 207-262.

Summers, R. and Heston, A. (1991), The Penn World Table (Mark 5): an Expanded Set of International Comparisons: 1950-1988, *The Quart. J. of Econ.*, vol. 106, pp. 327-368.

Sutherland, P. D. (1986), Europe and the principle of convergence, *Reg. Studies*, vol. 20, pp. 371-377.

SVIMEZ (1993), *I conti economici del Centro-Nord e del Mezzogiorno nel ventennio 1970-1989*, Bologna: Il Mulino.

Tagliacarne, G. (various years), Calcolo del reddito prodotto dal settore privato e dalla pubblica amministrazione nelle provincie e regioni d'Italia, *Moneta e Credito*.

Theil, H. (1967), *Economics and information theory*, Amsterdam: North-Holland.

Therkildsen, O. (1981), The Relationship between Economic Growth and Regional Inequality: A Critical Re-Appraisal, in Buhr, W. and Friedrich P. (eds), *Regional development under stagnation*, Baden Baden: Nomos.

Walsh, J.A. and O'Kelly, M.E. (1979), An Information Theoretic Approach to Measurement of Spatial Inequality, *Econ. and Soc. Rev.*, vol. 4, pp. 267-286.

Williamson, J. G. (1965), Regional inequality and the process of national development: a description of the patterns, *Econ. Dev. Cult. Change*, vol. 13, pp. 3-84.

Wolf, H. C. (1994), Growth Convergence Reconsidered, *Weltwirtschaftliches Archiv*, vol. 130, pp. 747-759.

5 Enhancing the Environmental Component of Structural Fund Regional Programmes: the Case of the East of Scotland

TONY JACKSON AND PETER ROBERTS

Introduction

Ecological Modernisation and Sustainability in the EU

Ecological modernisation challenges the conventional wisdom that there is a zero-sum trade-off between economic prosperity and environmental concern (Weale, 1992). In stating this, a suggestion is introduced that the achievement of a higher level of environmental protection is a precondition of (sustainable) long-term economic development. However, as Hajer (1996) argues, it is possible to identify a development pathway in the emergence of ecological modernisation discourse - moving from institutional learning, through ecological modernisation as a technocratic project, and eventually to a lasting shift in cultural politics. Generally in the UK and Western Europe, the first stage of this pathway has been delivered, and most European Union (EU) economies are now entering the second stage.

This application of ecological modernisation can be seen as an 'environment-as-technology' discourse coalition emerging from interaction between popular tendencies and technological interests in the political debates over the environment within industrialised market economies. As Jachtenfuchs and Huber (1993) have illustrated in their depiction of the EU response to radiative forcing, environmental policy concerns initially championed by radical green movements have largely been subsumed over the past decade into EU programmes designed to maintain the momentum of integration. A similar process   of cultural policy 'absorption' can be discerned in the development of environmental management standards within the EU (Franke and Waetzold, 1996).

In terms of Beck's critique of an emergent 'risk society' (Beck, 1992 and 1995), ecological modernisation can be seen as symptomatic of a tendency evident in society to replace the system of rewards traditionally based on ownership of the means of production in an industrial society with a far more random and arbitrary system of penalties generated by the risks produced by the technological processes of the modern industrial nexus.   Applying Hajer's terminology, within this critique ecological modernisation is view as a technocratic project in which the environmental concerns displayed by policy-makers, experts and scientists simply camouflage the crux of the modern problem, which is that technology itself is out of control. Proof of this is seen as evident in the risks to which those who control technological processes are prepared to subject society. Ecological modernisation becomes the means by which the hierarchical and centralist tendencies of technocracy apply the processes of modernisation to one of the remaining threats to the continuity of technocracy as the dominant political process (Welford, 1997).

Hajer's two alternative interpretations of the concept provide less provocative explanations of the processes of discourse.   Viewing ecological modernisation as institutional learning suggests that the ecological crisis is primarily a conceptual problem for policy-makers, seeking to find ways in which to incorporate environmental concerns into their areas of responsibility (Jackson and Roberts, 1997). Regarding the process as a matter of cultural politics, setting out the relationships between nature, society and technology, the agenda chosen by policy-makers offers the means to select suitable areas of debate.   Public discussions can be channelled in ways which pre-determine appropriate societal solutions, as evinced by the creation of sustainability fora and Local Agenda 21 programmes. This process of discourse can also be seen in the growing sophistication and coverage of the debate in the EU on the achievement of environmental goals through a range of policy instruments.

Whatever the primary motivation, the development of this form of environmental discourse is evident from the mid-1980s.   Zero-sum arguments rejecting economic growth in favour of conservation of the environment began to be replaced by positive-sum rationalisation of the need for sustainable growth to meet economic and social as well as environmental objectives. Some of the key documents in this process were the Brundtland Report (WCED, 1987), the World Bank Report on Development and the Environment (World Bank, 1992), subsequent output from the UN Conference on Environment and Development of the EU Fifth Environmental Action Plan (CEC, 1993). Jacobs (1991) offers a critique of the intellectual rationalisation under-pinning the creation of this discourse coalition, while von Weizsaecker et al (1997) offer a practical set of examples of its implications, to the point of sub-titling their book 'the new report to the Club of Rome'.

The widespread adoption of the concept of sustainability has succeeded in moving the environmental debate away from somewhat arid Pigovian partial equilibrium concerns about static inefficiency losses and appropriate regulatory, technical and fiscal instruments of control, which in themselves were reflective of a technological approach to environmental issues. Given an increasing awareness of limits to growth (compare Meadows et al, 1972, with Meadows et al, 1992), attention has now turned to steady-state resource dynamics and the means of attaining optimal sustainable growth paths. The WCC definition of sustainability is currently seen as an operational equivalent to the Brundtland concern with inter-generational equity of access.

> sustainable development means improving the quality of life while living within the carrying capacity of supporting ecosystems (WCU, UNEP, WWF, 1991).

This chapter seeks to identify and analyse the evidence of ecological modernisation within the mechanisms through which the EU promotes spatial cohesion and convergence within its boundaries. This is most readily identifiable in the instruments which have been developed to deliver coherent spatial patterns of assistance under the EU Structural Funds: the Single Programming Documents (SPDs). We have chosen a peripheral region, the East of Scotland, which contains a number of environmental issues pertinent to this discourse, as an example of current practice in this respect. The policy-making agencies within this areas are themselves advocates of sustainable development as a means of promoting the growth of the local economies within the East of Scotland (Eastern Scotland European Partnership, 1998).

This chapter starts by reviewing certain key aspects of the theory of ecological modernisation and, in particular, considers the contribution of economic theory to a clearer understanding of the new agenda. In this respect the methodology adopted to determine a rational allocation of resources for abating radiative forcing (Bruce et al, 1996) provides a classic test of Beck and Hajer's arguments. It is evident that deterministic or stochastic optimisation techniques applied to catastrophic uncertainties result in serious fallacies of composition when determining environmental priorities.

The following section offers a review of the current status of EU environmental policy, before examining the means by which this is translated into spatial initiatives on a systematic basis. This provides some insight into the extent to which the fragile regional and local discourse coalition established by sustainable development will survive any failure of ecological modernisation to tackle the spatial risks associated with continued economic growth. The East of Scotland offers a test-bed for exploring whether the social fragmentation contingent on a failure of the new paradigm to deliver both a better and more equitable quality of life can be avoided.

Theoretical Foundations of Ecological Modernisation

In examining the contribution of ecological modernisation to developing a viable pathway for the simultaneous achievement of economic growth and environmental enhancement, it is possible to identify a number of theoretical foundations. Three aspects of theory are of particular importance:

- ecological modernisation as a concept in theoretical contributions to environmental sociology;
- social science theory regarding the roles and organisational structures of institutional development;
- the insights provided by economic theory into optimal growth paths.

The first of these, the contribution to environmental sociology, is concerned with restoring balance in a social order that has become detached from its local roots. To restore the balance between social and environmental concerns, Mol (1995) suggests that a re-embedding should take place, with the institutionalisation of ecology in the social practices and institutions of production and consumption. In this area of debate, the reconstruction of a direct link between social choice, political structures, the economic sphere and ecology is seen as a priority.

In order to deliver the changes that are suggested in the first aspect of theory, a second consideration is the requirement for institutional development. Much of the work in this field has been stimulated by the ideas developed by Joseph Huber. His central argument is that organisational structures delimit and determine the application of technology to the resolution of ecological problems. It follows that the creation and implementation of appropriate institutions is a precondition for the achievement of a balanced economy, in which economic and environmental goals can be achieved within a common framework (Huber, 1983). This new framework provides a basis for the ecological transition of the industrial system and offers a response to the colonisation of nature by the industrial sphere which has been dominant for the past two hundred years.

The third important area of theory is an understanding of the foundations of ecological modernisation derived from the formal modelling of growth paths. It can be shown that under standard assumptions governing optimal growth paths, welfare growth is a function of the sustainable environmental stock and the proportion of reproducible capital assets needing to be switched from consumption to abatement of pollution (Mabey et al, 1997). 'Steady state' growth paths are sensitive to assumptions about critical environmental resource constraints. Given that our current position on such a path involves sub-optimal levels of abatement, in the absence of technological innovation the only dynamic pathway which does not violate sustainability requires the application of additional investment in abatement to a finite stock of environmental assets. This will ultimately generate diminishing returns to investment in abatement, calling into question the long-term viability of such a pathway. Only with the introduction of technological innovation, decoupling economic activity at least partially from environmental degradation, can a pathway be developed which offsets diminishing returns to investment in abatement and provides the possibility of sustainability.

The Effect of EU Environmental Policy on Ecological Modernisation

The allocative implications of such inter-generational dynamics are more restrictive than is generally supposed, especially given the further sustainability constraints imposed by Article 130r of the Treaty of Union, which requires that EU policy on the environment shall aim at a high level of protection and:

> be based on the precautionary principle and on the principles that preventative action should be taken, that environmental damage should as a priority be rectified at source, and that the polluter should pay.

As a recent review of the scope for abatement with respect to radiative forcing acknowledges:

> The precautionary principle ... can be broadly read as a simple statement advocating risk-averse decision-making due to the existence of hard uncertainty about irreversible and potentially catastrophic outcomes (Mabey et al, 1997, p220).

Removal of the assumption of risk neutrality common to stochastic versions of the standard dynamics model reinforces the case for technological innovation. Mabey et al (1997) demonstrates that in dealing with radiative forcing, risk neutrality over-values the benefits of wait-and-see strategies and under-values the return to risk-averse pro-active investments which may involve real abatement costs additional to 'no-regrets'.

Given the existence of hard uncertainty for which stochastic parametrics are inappropriate in dealing with abatement of radiative forcing, the 1996 IPCC report (Bruce et al, 1996) indicates the potentially high returns to investments in learning. These can both establish more clearly the variance of risk and also provide the means of identifying initiatives which fulfil the objective of modifying or hedging against the possible range of outcomes. Together with the deterministic case for technological innovation, the presence of uncertainty in the face of irreversibility and potentially catastrophic outcomes provides a powerful case for promoting ecological modernisation as institutional learning.

However, this process of institutional learning, at least in the case of the EU's policy mechanisms, has been subject to a number of delays and disruptions. These difficulties have their origins, as might be expected, in the multitude of roles that the EU attempts to perform. Of particular importance in the context of this paper is the dislocation that has existed between policies for economic growth and for environmental enhancement (Roberts, 1996). Many of the traditional policy objectives of the EU have been principally concerned with the promotion of economic growth, and it is only recently that environmental concerns have been promoted alongside this traditional agenda. As will be seen below, the latest EU environmental action programme (CEC, 1993) has been marked by a change in stance and an improvement in the means for simultaneous implementation of economic and environmental programmes.

The EU Environmental Action Programmes

The Fifth Action Programme on the Environment (EAP), which runs from 1993 to 2000, is entitled 'Towards Sustainability' (CEC, 1993). Both in the objectives it supports and in the modes of delivery it advocates it is strongly coloured by the ecological modernisation paradigm. Acknowledging the continued need for industrial development, it accepts as a primary EU objective the creation of conditions for a strong innovative EU industrial sector. Long term strategies to promote economic competitiveness within world markets for EU producers are supported, and the EAP envisages the creation of conditions which will allow the EU industrial sector to use its expertise in environmental technology and management practice to realise these. In contrast to a prescriptive, top-down approach towards industry adopted by previous programmes, realisation of this EAP rests on an acceptance that environmental quality and economic growth are mutually dependent, and reliant on a partnership approach between the principal actors:

> Under this programme the dual approach of high environmental standards combined with positive incentives to even better performance should be applied in a co-ordinated manner to the different points in the 'research-process-production-marketing-used-disposal' chain (CEC, 1993, p28).

Combined with this positive-sum approach towards objectives is a desire to apply instruments for its delivery which are more market-driven and self-regulatory.

Applying the terminology of the Commission, environmental measures meeting these requirements are seen as consisting of a mixture of 'vertical' initiatives, which focus on specific sectoral projects to deliver environmental improvements at all spatial levels within the EU; and 'horizontal' initiatives, tying together the separate elements of all EU programmes at a given spatial level and giving them an appropriate environmental aspect. Given the consensual approach advanced by the EAP and the limited funds available for specific environmental initiatives, horizontal measures are clearly of great importance in delivering the objectives of the current EAP.

Horizontal environmental measures first came to prominence in the Fourth EAP (1988-92). The Single European Act (SEA) established environmental protection as an obligatory component of all EU policies. For the structural programmes related to agriculture, regional and social development (EAGGF, ERDF, ESF), this was to be realised by a process termed 'external integration' or each initiative with EU environmental policies (as distinct from the 'internal integration' expected of EU environmental policies themselves).   Such requirements encouraged the evolution of disbursement arrangements for these traditional funds into the current partnership-based managed programmes, which now deliver predominately soft forms of assistance such as business support, training and education, rather than infrastructural investments (Roberts and Hart, 1997).   However, as Seamark (1996) has demonstrated, many of the environmental appraisals conducted as part of the process of preparing and approving the 1994-1996 round of Structural Fund programmes were lacking in consistency, detail and an explicit means for the implementation of environmental objectives.   The new programmes, for the period 1997-1999, provided an opportunity to improve on this somewhat dismal performance.

Regional Assistance Delivered Through the EU Structural Funds

The current programme for the Structural Funds was agreed in 1993, and runs from 1994 to 1999.   There are six specific categories of assistance, grouped under five objectives, three of which (Objectives 1,2 and 5b) are spatial in nature and require the establishment of programmes of regional assistance.   Area programmes designed for this purpose are drafted and delivered on a partnership basis between the Commission and consortium of public, private and voluntary sector representatives within member states led by an implementing body representing the national government.   The partnership consists of representatives from regionally-based bodies with development responsibilities for the area, including local government, enterprise and training agencies, private sector organisations, further and higher education and the voluntary sector.   These partners draw on local, regional and national funding sources in addition to those available through the programme to put together proposals and make bids for funding projects which meet the criteria set out in the programme.   Private sector financial support for the programme is also sought.

The 1993 regulations established a Single Programming Document (SPD), which allows the operational programme for delivering structural assistance to an area to be combined with the regional development plan on which the priorities and measures are based. This has reduced the time lags involved in agreeing and delivering a programme.   The development plan sets out the rationale behind the structural programme of regional assistance. The SPD allows this to be linked directly to a set of priorities identified for realising the plan, and a set of measures for meeting each of these priorities. In this way, the programme establishes the rules for a bidding process to fund proposals designed to realise the plan.

In Scotland, a Programme Monitoring Committee (PMC) chaired by a representative of the national government scores bids and selects proposals, based both on the 'vertical' priorities identified, and also on the 'horizontal' priorities which must automatically be included, including environmental and socio-economic objectives. The PMC operates alongside a Programme Management Committee which represents the interests of the regional partners, reviewing progress and making arrangements for evaluation. The Scottish Office Development Department is the implementing authority for all Scottish regional programmes of assistance. The day-to-day management of a programme is undertaken by an arm's length body knows as a Programme Executive.

Within the UK context, this process is important for two distinct reasons. Firstly, it provides an operational basis for the exercise of regional assistance under a system which (if the territorial executives for Scotland, Wales and Northern Ireland are discounted) experienced no other significant form of coherent regional planning for many years. Roberts and Hart (1996) regard the EU regional programmes of structural assistance as:

> one of the factors which have helped to stimulate a resurgence of regional planning in the UK (p3).

The increasing need for inputs at a regional level undoubtedly contributed to the decision to create Government Offices for the Regions in England.

Secondly, the EU regional programmes also offer an opportunity not otherwise available in the UK to test the viability of delivering on an integrated spatial basis a set of environment measures designed to implement sustainable development at a strategic level. The fifth EAP specifically identifies actors and activities that create depletion of resources and produce other adverse environmental impacts. It also suggests that these 'target groups' (a term derived from the Dutch National Environmental Policy Plan MHPPEN (1994) on which this EAP draws heavily), can be drawn into co-operative plans delivered to greatest effect at a strategic level, through the application of vertical and horizontal environmental measures focused on spatial initiatives within a region.

Disregarding the limited funding available through the Community Financial Initiative for the Environment (LIFE), for member states ineligible to draw on the Cohesion Fund the principal means by which the objectives of the fifth EAP can be translated into practical measures is through the Structural Funds. Indeed, the fifth EAP sees reliance on delivery of much of its programme via the Structural Funds as a means of inducing changes on the part of regional planners and developers towards the use of environmental resources, so that structural development programmes are put together in ways which incorporate ecological modernisation as the basis for continued socio-economic improvement. As a former Director-General of Directorate XI has observed, underlying the strategy of the fifth EAP is the assumption that the general objective of sustainable development can only be achieved through a joint effort by all parties in the form of a partnership (Brinkhorst and Klatte, 1993, p73). Hanf (1996) argues that the approach currently being pursued reflects a desire to identify the appropriate regulatory space for delivery of environmental initiatives.

Although the fifth EAP acknowledges that the realisation of long term goals requires short to medium term programmes capable of being monitored and evaluated, it studiously avoids identifying a comprehensive list of measures. In part, this is a reaction against the shopping list approaches characteristic of previous EAPs. It is also a conscious reflection of a desire to facilitate rather than control the process of adaptation required, because:

> it is not feasible to adopt a Directive or Regulation which says 'Thou shalt act in a sustainable manner' (CEC, 1993, p64).

As a consequence, the emphasis has switched from top-down regulation to promotion of measures at an appropriate level and through co-operative actions to deliver environmental initiatives as part of other EU programmes. These include, apart from the structural programmes, various marked-based initiatives which seek to internalise environmental externalities. This arm's-length approach, despite the enhanced powers acquired over environmental policy by the EU following the SEA and Treaty of Union, also reflects the political realities imposed on the EU by high levels of unemployment and slow economic growth.

The Eastern Scotland Programme Area (ESPA)

An examination of the environmental input in a recently drafted SPD provides some insight into the effectiveness of the Structural Funds as a mechanism for 're-ordering regulatory space' to deliver the fifth EAP and promote ecological modernisation. Objective 2 programmes run for half the full period covered by the current EU Structural Funds programme, allowing some flexibility within the programme under this objective. The Eastern Scotland Programme Area (ESPA) has just embarked on a second SPS (CEC, 1997), covering the period 1997-99, having been advised in January 1996 that the status of the area would be maintained for the remainder of the programme period.

EASTERN SCOTLAND OBJECTIVE 2 AREAS

Figure 5.1 The Eastern Scotland Objective 2 Programme Area

As Figure 5.1 indicates, the ESPA is not territorially contiguous. It consists of parts of four former regions (Central, Fife, Lothians and Tayside), and includes all of seven travel-to-work areas (Arbroath, Alloa, Bathgate, Dundee, Dunfermline, Falkirk, Kircaldy) plus parts of two others (Edinburgh and Stirling). Together, these eligible areas span most of the estuary and Firth of Forth, and part of the northern side of the Firth of Tay.

For the purposes of EU structural assistance under Objective 2, the predominant feature linking these communities is a chronic loss of employment in production and manufacturing. Objective 2 focuses on the conversion of regions, frontier regions or parts of regions including employment areas and urban communities seriously affected by industrial decline. The regional and social conversion plan for such areas is intended to establish a framework to generate additional sources of employment compatible with socio-economic and environmental requirements of the area. Additional severe job losses created by closure of deep-mined coal pits and the run-down of naval dockyards in Eastern Scotland are covered by the separate RECHAR and RENEVAL programmes of assistance to the affected communities.

The ESPA has a high concentration of electronics and instrument engineering, located mainly in its two new towns, Glenrothes and Livingston, and to a lesser extent also in Dunfermline and in Dundee. However, buoyant demand for electronics and steadily rising employment in this sector has not offset far larger job losses over recent decades in traditional manufacturing and production activities, such as textiles, brewing and distilling, chemicals and mechanical engineering.

Manufacturing employment in the ESPA fell by just over 25 per cent between 1984 and 1993, although at 22.1 per cent of the workforce, it still remains well above the average for Scotland (16.1 per cent) and for GB (18.1 per cent). Some distortion of the statistical base is created by the exclusion of most of the City of Edinburgh, the largest community in the catchment area, but the ESPA clearly suffers from rates of unemployment significantly higher than the Scottish and GB averages. The occupational profile reveals a labour force with lower than average skills levels and self-employment, reinforced by lower than average business birth rate. Together these are classic symptoms of areas overly dependent on traditional sources of manufacturing employment.

Aside from its electronics sector, the areas shows other potential for technologically-driven growth. It is served by seven universities, four major research institutes and fifteen further education colleges. Some 700 of the 2,500 Scottish companies (28 per cent) identified in a recent survey as having research and development activity (CEC, 1997, p34) are located in the ESPA. Some offsetting weaknesses have also been identified, including a low research and development involvement by the private sector, limited integration between further and higher education and business, and a poor rate of high-tech business start-ups.

Environmental Inputs in the ESPA Strategic Plan

The SPD for the area identifies four strategic priorities in its regional regeneration strategy:

- development of a dynamic indigenous SME base
- tourism
- locally-based initiatives
- technology and innovation

Locally-based initiatives refer to community schemes designed to tackle socio-economic deprivation, while the other three priorities focus on aspects of indigenous business growth, identifying two sectors, tourism and technological innovation, as offering scope for assisted development.

In addition to these 'vertical' priorities, any SPD is required to address common 'horizontal' priorities laid down for all the Structural Funds which are currently grouped under the headings of:

* the environment
* equal opportunities
* human resource development

It is possible to regard these 'horizontal' objectives taken together as fulfilling the intra- and inter-generational equity and environmental requirements of sustainable development, although in practice they are regarded as setting separate 'horizontal' requirements for the plan. The ESPA programming document contains a total of twelve measures aimed at meeting its four strategic priorities, SME support, tourism, local initiatives and technology and innovation. Some of these measures also directly address at least two of the 'horizontal' priorities. Promoting growth of small and medium enterprises (SMEs) (Priority 1) and emphasising community economic development under local initiatives (Priority 3) are seen as contributing towards realisation of equal opportunities, while each of the priorities has attached to it a human resource development training measure.

By contrast, although efforts to increase the technological capacity of SMEs might be seen as providing opportunities for ecological modernisation, the SPD itself admits that despite the Commission's view that the environmental dimension should be treated at both the 'horizontal' and 'vertical' level the document:

> does not specifically address the environmental and sustainable development at a vertical level (CEC, 1997, p78).

The justification offered for this omission is that environmental issues are sufficiently incorporated in the programme through the 'horizontal' linkages included in all the measures and through the Partnership's management and monitoring structures, which will promote sustainability.

What this means in practice is that the SPD included only a token quantification of environmental outputs and impacts. Amongst the 85 specific priority targets set for the programme, which cover the economic impact, intermediate outputs and physical activity to be generated by each measure, Figure 5.2 indicates that there are only four environmental ones. These include 20 ISO and 10 EMAS registrations (which are only recorded as targets in the individual measures and do not figure in the aggregated priority targets), the reclamation of 45 hectares of derelict and contaminated land, and environmental improvements to 50 hectares of land.

It is premature to pass judgement on the effectiveness of this approach towards integrating the fifth EAP into the current round of regional assistance. Nevertheless, the document offers some pointers, which suggest that the interpretation given to sustainable development by the partnership is in itself compatible with the paradigm of ecological modernisation, and is much more limited in scope and imagination than the approach taken by some other comparable member states. This is evident, for example, in the justification offered for the approach taken by the SPD to the 'environmental' dimension of sustainable development.

> Respect for the environment and the extent to which the project promotes environmental sustainability will be key elements of the project selection criteria for each and every measure. Key to the new environmental impetus for the SPD 1997-1999 in promoting greater awareness of the environment, ensuring compliance with national and EU legislation and promoting environmental sustainability will be the active participation of the two competent environmental authorities, i.e. Scottish Natural Heritage and the Scottish Environmental Protection Agency at all levels of the management and implementation of the new SPD (CEC, 1997, p78).

This passage appears to view the issue largely as a matter of applying specific measures designed to boost regional employment a traditional watching brief towards environmental protection, exercised by central government's appropriate representatives. This restricted definition of the scope of the environmental dimension of sustainable development is not untypical of Structural Funds regional programmes.

No of targets under each priority*

Type of target	SME support	Tourism	Local Initiatives	Technology & Innovation
Economic impact				
Employment	6	6	5	6
Output	3	3		3
Income		1	1	
Intermediate output				
Businesses	1		1	
Products/processes	1			2
Markets	1			
Trainees	3		1	
Skills & qualifics	1			
Reduced unemployment			1	
Env standards	(2)**			
Visitors		4		
Physical activities				
Financial assistance	1			
Counselling	1			
Enquiries	1			
Premises	2		2	5
Sites	1			
Roads	2			
Attractions		2		
Community Initiatives			1	
Contam/derelict land	1			
Env Improvements			1	
Training/qualifics		4	4	3
Training facilities			2	
Marketing		1		
Educations Links				1
Total	25	21	19	20

Figure 5.2 Targets set for Eastern Scotland SPD, 1997-1998
Source: CEC 1997
Environmental targets in italics
** Identified as target under a specific measure but not in overall summary of targets under priorities.

Appraisal criteria	Priority measure*											
	1.1	1.2	1.3	1.4	2.1	2.2	2.3	3.1	3.2	4.1	4.2	4.3
Minimise use of non-renewable resources	=	+	=	+	=	=	=	=	=	+	=	=
Environmentally-sound use and management of hazardous/polluting substances and wastes	=	+	=	=	=	=	=	=	=	+	=	=
Use renewable resources within the limits of capacity for regeneration	=	=	=	=	=	=	=	=	=	+?	=	=
Maintain and improve quality of wildlife, habitats and landscapes	=	=	=	+?	=	+/-?	=	=	=	=	=	
Maintain and improve quality of soils and water resources	=	=	=	+?	=	-?	=	+	=	=	=	=
Maintain and improve quality of historic and cultural resources	=	=	=	+?	=	+/-?	=	=	=	=	=	=
Maintain and improve local environmental quality	=	=	=	+	=	+	=	+	=	+?	=	=
Protection of global atmosphere	=	=	=	=	=	=	=	+	=	+?	=	=
Develop environmental awareness, education and training	=	+	+?	=	=	+?	+?	=	+	+	=	+
Promote public participation in decisions involving sustainable development	=	=	=	=	=	=	=	+	=	=	=	=
Overall assessment	=	+	=	=	=	=	=	+	=	+	=	=

Key:

+   Potential for significant beneficial impact

-     Potential for significant adverse impact

"   No relationship or significant impact

?   Likelihood of beneficial or adverse impact uncertain

+/-?   Impact direction uncertain

Priority measures:

1.1   Support for business start-ups

1.2   Support for SME growth and development

1.3   Assisting HRD for SMEs and employment growth areas

1.4   Investments in productive facilities

2.1   Support of SMEs in tourism and cultural industries

2.2   Support for strategic tourism development

2.3   Promotion of targeted tourism skills development and employment support

3.1   Community economic development (ERDF)

3.2   Community economic development (ESF)

4.1   Increasing technological capacity of SMEs

4.2   Provision of facilities to strengthen local R & D, technology and transfer, and HRD systems for SMEs

4.3   Developing skills in technology and innovation geared to needs of SMEs

Figure 5.3 Matrix used for appraisal criteria for programme measures in ESPA SPD, 1997-1999 (Source CEC 1997)

Environmental Appraisal in the ESPA SPD

Figure 5.3 reproduces the matrix used by the SPD to illustrate how environmental concerns are currently taken into account within the document. It involves the application as appraisal criteria of a Scottish Natural Heritage (SNH) checklist created for strategic environmental assessment of European Structural Fund Programmes. This offers further confirmation that the full implications of strategic planning and assessment for sustainability have not been absorbed. For example, the row dealing with 'protection of the global atmosphere' suggests that despite an outlay which is designed to make an appreciable impact on local activity, none of the measures should produce a detrimental environmental effect of any significance.

Reference to the environmental review which precedes this table reveals that the ESPA has four coal-fired power stations and two major petro-chemical sites. Air-borne emissions from these sources are already a serious problem:

> In 1989, the highest levels of sulphur dioxide emissions in Scotland were recorded within the programme area, with over 20,000 tonnes of SO_2 emitted in an $800km^2$ area between Stirling and Dunfermline (CEC, 1997, p50).

The environmental review goes on to observe that in Edinburgh in 1994 the proposed UK National Air Quality Strategy standard for sulphur dioxide was exceeded on thirty occasions. This national standard reflects the cut in emissions of sulphur dioxide and nitrogen oxides agreed by the UK Government in 1994 in response to the 1988 EU Large Combustion Plants Directive. It is far less stringent than the EC acidification proposals currently under consideration, which aim to cut acid gas emissions in the EU by 80 per cent by 2010. The Electricity Association has been quoted as saying that these proposals would limit fossil-fuel electricity generation for the UK "to the single option of gas-fired power stations with additional emission controls", that it would force the closure of all coal-fired stations, and that other industries covered by the directive, such as steel and chemicals, "may relocate operations abroad rather than upgrade existing plants" (Sunday Times, 1997).

The matter of what is an appropriate level of emissions for such plants may not be for the SDP to determine, but it is remarkable that the SPD seems oblivious of the implications of these issues for its targets, and appears to regard its own demands for energy as having no adverse consequences worth noting under this heading. By contrast, the Dutch approach (MHPPEN, 1988, 1994) is to encourage the setting of targets for reduction of emissions from power stations and other large plants through joint negotiations on a partnership basis between representatives of industry and the planning and development authorities at an appropriate spatial level, leaving it to regional or local initiatives to come up with proposals which contribute to the targets.

The coal-fired power stations in the area also pose a major problem related to the disposal of fly ash, much of which is currently dumped into slurry lagoons in the Forth estuary. Technology is available from establishments located within the ESPA to turn such waste into inputs for various other processes that could reduce demands on raw materials and counteract acidification of the soil. These processing opportunities provide good examples of the systems approach towards the management of environmental impacts, which includes the closing of material cycles advocated by supporters of the industrial ecology mode of ecological modernisation (Allenby and Richards, 1994; Ayres and Ayres, 1996). Specific proposals have recently been made for the Grangemouth petro-chemical complex to have a gas-fired combined heat and power plant and for a biomass power station to open at Westfield near Kircaldy, designed to burn chicken waste to generate ten megawatts of electricity per year.

It is therefore surprising that the SPD makes little effort to consider the energy efficiency aspects of its programme, how these might be met in environmentally-sustainable fashion, and how such initiatives could generate further business elsewhere. Other major issues are also ignored. Most of the appraisal criteria employed in the matrix of Figure 5.3 consist of environmental quality indicators as distinct from indicators of sustainability (Touche Ross, 1994, 1995). Their choice reflects the rural priorities of SNH, rather than the urban environmental problems afflicting many of these industrial communities. For example, the matrix fails to reveal the impact of the plan on the serious congestion and pollution around the crossings of the Forth and the approaches to Edinburgh created by ongoing traffic growth on the roads. The absence of an integrated transport management scheme for the area hardly assists. It is unlikely that fulfilment of the targets within the programme will ease this problem. Indeed, growing congestion may act as a serious constraint on delivery of some of the measures in certain locations.

Similarly, the ESPA has serious effluent discharge problems arising from both abandoned mines, and the general lack of secondary treatment for most sewerage discharges to seat. The commitment of the UK Government to meeting EU standards is likely to result in significant increases in water charges to local business, reducing their competitiveness, unless appropriate measures to limit discharges can be incorporated within any regional regeneration strategy.

As already suggested, sustainability could be seen as encompassing all three of the horizontal priorities: initiatives which simultaneously promote improvements in environmental quality and in socio-economic conditions within the area. This requires a far more complete and integrated approach towards sustainable development than a checklist designed to see whether a programme chosen on other grounds complies with certain environmental precepts. Such a systematic strategic approach towards environmental assessment is advocated in the European Sustainable Cities Report (CEC, 1996). This stresses the importance of viewing urban and rural communities as complex ecosystems, dealing with flows of environmental resources which are susceptible to indirect and induced as well as direct impacts.

Some recognition of the shortcomings of the approach currently adopted by the Partnership is evident from a careful reading of the SPD. The current EU Structural Funds programme places particular emphasis on appraisal procedures (Bachtler and Michie, 1995). These are seen as consisting of a three-stage process, covering prior appraisal or proposals, monitoring and interim assessment of delivery, and subsequent evaluation of impact. The present regulations require all SPDs to include:

- an appraisal of the current environmental situation;
- an assessment of the impact on the environment caused by the plan;
- the designation of an environmental authority for the region.

In reviewing the lessons learned from the previous round, the current SPD acknowledges the need to improve its approach towards appraisal of environmental issues. It reveals that a study of environmental aspects of the previous round concluded that it lacked "a clear focus for what was required in terms of project development" (CEC, 1997, p89) to deliver the sustainability requirements expected by the Commission. The same part of the document (Ch.3: Regional Regeneration Strategy Consolidation and Refocusing) lists the following issues which need to be given a more strategic consideration:

- transport planning
- energy efficiency
- forestry and biomass
- waste minimisation

- brownfield sites
- public awareness and training in environmental skills.

The current SPD undertakes to address these issues through further work, which will focus on:

- a more comprehensive environmental baseline review of the ESPA;
- revision of its project appraisal system to include environmental assessments at an early stage;
- administrative charges to include further environmental expertise within the partnership;
- development and promotion of good practice environmental guides for partners, plus promotion of eco-business;
- the development of monitoring criteria to assess the environmental content of the Programme.

These are welcome improvements which acknowledge the reservations expressed by Bachtler and Michie (1995) about the capacity of regional partnerships to incorporate effective environmental appraisals of their impacts in accordance with the structural fund regulations. Nevertheless, they do not wholly address the lack of a clear strategic appreciation of the meaning and requirements of planning and managing sustainable development, a skill which is more evident within the planning processes of Fife Council (Fife Council, 1996 and 1997, Jackson and Roberts, 1997). The appointment in 1995 of Fife Council as the lead partner combined with merging of the four sub-regional programmes used for the 1994-96 phase of the ESPA into a single programme with a Programme Executive based in Fife, should in itself assist in this learning process, as will the addition of the Scottish Environmental Protection Agency (SEPA) to the Partnership.

Mention should also be made of a significant initiative being undertaken on a partnership basis within the ESPA, termed the Forth Forum.

The Forum is currently working to develop an integrated management strategy for the whole of the Forth estuary by 1998. Established in 1993, it owes its origins to the UNCED Conference and UK adherence to the Concordat on Biological Diversity signed in 1992. This recognises the importance of major river estuaries and the increasing pressure on them from economic development. Led by SNH and SEPA, the partnership includes a wide range of representatives from the public and private sectors, and is undertaking its work through ten topic groups:

- economic development
- coastal and marine pollution
- nature conservation
- tourism and recreation
- coastal defence
- landscape and amenity
- awareness and education
- fisheries
- built and archaeological heritage
- information and research

Work commissioned for the economic development group seeks to identify current and future challenges to the sustainable development of the project area, the potential for removing or ameliorating such challenges, and the priorities and mechanisms for effectively addressing these. On completion, the Forth Forum Management Strategy should provide a much needed framework for establishing the parameters within which any regeneration strategy under the Structural Funds must operate in order to promote sustainable development in this part of the ESPA. A similar initiative for the Tay Estuary is in its early stages.

At the same time, the Scottish voluntary sector is also making an important contribution to the meaning and practice of sustainable development. The discussion paper, 'Towards a Sustainable Scotland' produced by Friends of the Earth Scotland (1996), offers two alternative methods of assessing strategic environmental impacts. One is the notion of environmental space, developed by the Centre for Human Ecology at the University of Edinburgh. This calculates the current overall pattern of resource consumption and availability within the Scottish economy, and provides an indication of changes required to meet future sustainable targets.

From this, it is possible to assess the scale of the task to modify both consumption and production patterns over the next decades to achieve a more sustainable use of resources. Although not a precise planning tool, such work can highlight the key sustainability parameters requiring attention within any economic regeneration plans. The second approach, developed by the University of Stirling as an offshoot of the Scottish Office Sustainable Systems Project, provides an 'Index of Sustainable Economic Welfare' (ISEW), modifying measures of Scottish Gross Domestic Product to include social and environmental inputs. This has the effect of reducing the apparent rate of economic growth, and could be used to provide more apposite indicators of the impact of regional assistance, which incorporate the current horizontal priorities.

Within the Scottish Office itself, work has been undertaken on ways of utilising the Scottish input-output tables to provide estimates both of the impacts of changes in economic activity on the environment, and of the impact of changes in environmental parameters on the economy (Steedman and Vickerman, 1993; Alexander and McNicholl, 1995). Although at a preliminary stage, all these exercises in quantifying environmental impacts of economic change represent an essential component in the process of creating an effective regional management structure capable of identifying and promoting spatially-based strategies and programmes which can deliver ecological modernisation.

Conclusions

In attempting to assess the progress being made in delivering ecological modernisation via the regional partnerships of the Structural Fund, the conclusions to be draws from an examination of the ESPA SPD confirm Hanf's point about the need for re-ordering of regulatory space for delivering environmental initiatives (Hanf, 1996). Hanf suggests that EU environmental policy is prompting the creation of a new equilibrium of regulatory space governing the relations between government and economic actors, better able to promote the objectives of ecological modernisation.

Such a re-ordering is further advanced in some EU member states than it is in the UK. Current governmental arrangements within the UK do not at present provide an adequate basis on which to establish the strategic sub-national framework necessary for the creation and development of a sustainable development planning framework. The regional partnerships created for EU structural assistance, which at present cover 85 per cent of the Scottish population, offer a vehicle for testing the alternative possible frameworks. A position paper drafted by the Convention of Scottish Local Authorities on the future of the Structural Funds after 1999 recognises this point (COSLA, 1997). It argues that the principles of sustainability should be fully incorporated into the EU Structural Funds programme, and that support should only be given to programmes and projects which:

- maintain the overall quality of life;
- maintain continuing access to natural resources;
- avoid lasting environmental damage;

- meet the needs of the present without compromising the ability of future generations to meet their own needs;

In particular, 'all programmes should be subject to a full sustainable development assessment of their policy objectives and all projects subject to a rigorous environmental impact assessment' (p14).

Amongst the regional and local authority partners within the ESPA, considerable progress has already been made in designing, developing, delivering and measuring the impact of sustainability initiatives. Jackson and Roberts (1997) have demonstrated for one of these councils how the implementation of sustainability policies has entailed changes to traditional management structures in local government, the use of environmental management systems, and the development of performance measures in the form of sustainability indicators. Each of the ESPA local authorities has a strong commitment to Local Agenda 21 (LA21). Part of this involves the establishment of LA21 fora, providing local community interests with the means to share in developing approaches to sustainability that mesh with the fifth EU EAP. Here the emphasis is on the participatory elements of sustainability - consensus building, empowerment and strategic management (Bruder, 1997) - as a counter-balance to the policy-orientated functional systems of management and monitoring required in local authority programmes.

A similar need is recognised south of the border, where the regional framework is even less well developed. In response to the discussion paper issued by the Department of the Environment, Transport and the Regions on the proposed creation of Regional Development Agencies for the English regions, the Town and Country Planning Association responded by urging that these bodies be used to ensure sustainable regional planning and development (TCPA, 1998):

> The need for an environmentally-driven sustainable approach to regional development is especially important in the older industrial regions where dereliction and other forms of environmental degradation act as a major impediment to economic and social development. In addition, the desirability or working with the needs of the environment in mind is recognised as coincident with the maximisation of economic potential; the market for environmental goods and services is growing rapidly and outstrips most other markets (pp6-7).

Roberts and Chan (1997) illustrate how regional and local authority initiatives, such as those pursued by the Association of Greater Manchester

Authorities, can provide a model for delivering strategies of sustainable development based on ecological modernisation. This experience provides evidence that supports an approach that emphasises development and monitoring of long-term strategies for sustainable development on a spatial and cross-sectoral basis, involving a partnership approach which sets environmental targets and supports local and regional initiatives to meet these. The present Dutch approach meets many of these requirements. It seeks to establish at appropriate levels a comprehensive set of voluntary partnerships working together to deliver programmes capable of meeting specific environmental targets:

> The final responsibility for implementing policy lies with the target groups. In an ideal situation the target groups would modify their behaviour so as to realise sustainable development ... Implementation will be pursued and instruments chosen so as to achieve a better coincidence of collective and individual interests ... The scope given to target groups, provinces and municipalities and intermediary organisations, to make their won choices can be enlarged, with a clear framework (MHPPEN, 1994, p42).

The same source goes on to state that the targets set for such groups, which are an essential component of this approach should be fixed within an open planning process in consultation with each group, so as to meet the overall objectives of the plan. The target groups should accept responsibility for their own targets and determine by themselves how they will achieve these, subject to monitoring and reporting procedures. The fifth EAP has clearly been based on a belief that the Dutch approach is the most likely to deliver ecological modernisation.

Roberts (1994) has suggested that three themes are likely to dominate the agenda in relation to the design and implementation of strategies for sustainable development at a regional level:

- the search for forms of economic organisation that respect the environment and minimise the negative environmental consequences of development;

- the desirability of moving towards spatial forms and modes of social organisation that minimise the excessive use of resources and maximise environmental benefits;

- the desirability of meshing together sectoral and spatial elements to ensure the environmentally responsible and balanced planning and development of regions (p781).

Despite evidence from other member states that the programmes supported by the Structural Funds offer an opportunity of fulfilling at least part of this agenda, the example of the Eastern Scotland regional assistance programme illustrates that current regional programmes in the UK have made only limited progress in this respect. Failure to deliver a more focused and targeted set of strategic environmental initiatives in future SPDs will support Beck's contention that ecological modernisation is essentially a smokescreen to camouflage the growing environmental damage that technology is imposing on society.

References

Alexander, J. M. & McNicholl, I. H. (1995), Social accounting matrices - an extension of the Scottish 1989 input-output analysis, *Scottish Economic Bulletin*, 51, pp29-46.

Allenby, B. R. & Richards, D. J. (eds.) (1994), *The greening of industrial ecosystems*, Washington DC: National Academy Press.

Ayres, R. U. & Ayres, L. W. (1996), *Industrial Ecology: Towards Closing the Material Cycle*, Cheltenham: Edward Elgar.

Bachtler, J. & Michie, R. (1995), A New Era in EU Regional Policy Evaluation? The Appraisal of the Structural Funds, *Regional Studies,*29.8, 745-752.

Beck, U. (1992), *Risk society: Towards a new modernity*, London: Sage.

Beck, U. (1995), *Ecological policies in a age of risk*, Cambridge: Polity Press.

Brinkhorst, L. J. & Klatte, E. R. (1993), EC environmental policy: evolution and perspective, in Spaargarten G (ed) *International environmental policy*, The Hague: Als.

Bruce, J. P., Lee, H., Haites, E. F. (1996), *Economic and social dimensions of climate change, vol.3, Climate change 1995*, Intergovernmental Panel on Climate Change, Cambridge: Cambridge University Press.

Bruder, N. (1997), Lessons from environmental auditing for the development of local environmental policy, in Farthing S M (ed) *Evaluation of Local Environmental Policy*, Aldershot: Avebury.

Commission of the European Communities (1993), *Towards Sustainability: European Commission Programme of Policy and Action in Relation to the Environment*, Brussels: Commission of the European Communities.

Commission of the European Communities (1996), *European Sustainable Cities - Report of the Expert Group on the Urban Environment*, Brussels: Commission of the European Communities.

Commission of the European Communities (1997), *Eastern Scotland Single Programme Document 1997-99*, Brussels: Commission of the European Communities.

Convention of Scottish Local Authorities (1997), *Future of the Structural Funds interim position paper*, Edinburgh: Convention of Scottish Local Authorities.

Eastern Scotland European Partnership (1998), *The Sustainable Development Project: Consultative Report*, Dunfermline: Eastern Scotland European Partnership.

Fife Council (1996), *Sustainability policy*, Glenrothes: The Planning Service, Fife Council.

Fife Council (1997), *First sustainability action plan*, Glenrothes: The Planning Service, Fife Council.

Franke, J. F. & Waetzold, F. (1996), Voluntary initiatives and public intervention - the regulation of eco-auditing, in Leveque F. (ed) *Environmental policy in Europe: industry, competition and the policy process*, Cheltenham: Edward Elgar.

Friends of the Earth Scotland (1996), *Towards a Sustainable Scotland*, Edinburgh: Friends of the Earth Scotland.

Hajer, M. A. (1996), Ecological modernisation as cultural politics, in Lash, S., Szerszynski, B., Wynne, B. (eds) *Risk, environment and modernity: towards a new ecology*, London: Sage.

Hanf, K. (1996), Implementing international environmental policies. in Blowers, A. & Glasbergen, P. (eds) *Prospects for environmental change*, London: Arnold.

Huber, J. (1983), Die neue sozialen Bewegungn zwischen Konfrontation und Kooperation, in Fink U (ed) *Keine Angst vor Alternativen*, Freiburg.

Jachtenfuchs, M. & Huber, M. (1993), Institutional learning in the European Community: the response to the greenhouse effect, in Liefferink, J. D., Lowe, P. D., Mol, A. P. J. (eds) *European integration and environmental policy*, London: Belhaven Press.

Jackson, T. & Roberts, P. (1997), Greening the Fife economy: ecological modernisation as a pathway for local economic development, *Journal of Environmental Planning and Management*, 40.5, 617-633.

Jacobs, M. (1991), *The Green Economy*, London: Pluto Press.

Local Government Management Board (1994), *Local Agenda 21: principles and process, a strategic guide*, Luton: Local Government Management Board.

Mabey, N., Hall, S., Smith, C., Gupta, S., (1997), *Argument in the greenhouse: the international economics of controlling global warming*, London: Routledge.

Meadows, D. H., Meadows, D. L., Randers, J. and Belwens, W. (1972), *The Limits to Growth: a Report for the Club of Rome on the Predicament of Mankind*, London: Earth Island Press.

Meadows, D. H., Meadows, D. L., Randers, J. (1992), *Beyond the Limits*, London: Earthscan.

Ministry of Housing, Physical Planning and Environment (1988), *To choose or lose - National Environmental Policy Plan*, The Hague: Ministry of Housing, Physical Planning and Environment of the Netherlands.

Ministry of Housing, Physical Planning and Environment (1994), *The Netherlands' Environmental Policy Plan 2*, The Hague: Ministry of Housing, Physical Planning and Environment of the Netherlands.

Mol, A. P. J. (1995), *The Refinement of Production*, Utrecht: Van Arkel.

Roberts, P. (1994), Sustainable Regional Planning, *Regional Studies*,28.8, 781-787.

Roberts, P. (1996), European Spatial Planning and the Environment: Planning for Sustainable Development, *European* Environment, 77-84.

Roberts, P. and Hart, T. (1996), *Regional Strategy and Partnership in European Programmes: Experience in Four UK regions*, York: Joseph Rowntree Foundation.

Roberts, P. and Chan, R. C. K. (1997), A Tale of Two Regions: Strategic Planning for Sustainable Development in East and West, *International Planning Studies,*2.1, 45-62.

Roberts, P. and Hart, T. (1997), The Design and Implementation of European Programmes for Regional Development in the UK, in Bachtler, J. and Turok, I. (eds) *The Coherence of EU Regional Policy*, London: Jessica Kingsley.

Seamark, D. (1996), European funding and environmental appraisal, *Town and Country Planning,* 65, 340-345.

Steedman, J. & Vickerman, J. M. (1993), Road transport costs in Scottish Industry and the impact of a carbon/energy tax, *Scottish Economic Bulletin*, 47, 15-28.

Sunday Times (1997), UK fights EC pollution plan, *Sunday Times*, London, 3 August.

Town and Country Planning Assocation. (1997), *Regional development agencies: ensuring sustainable regional planning and development*, London: Town and Country Planning Assocation.

Touche Ross (1994), *Sustainability indicators research project: report of phase 1*, Luton: Local Government Management Board.

von Weizsaecker, E., Lovins, A. B., Lovins, L. H. (1997), *Factor four: doubling wealth - halving resource use: the new report of the Club of Rome*, London: Earthscan.

Weale, A. (1992), *The new politics of pollution*, Manchester: Manchester University Press.

Welford, R. (1997), *Hijacking environmentalism: corporate responses to sustainable development*, London: Earthscan.

World Bank (1992), *Development and the Environment: World Development Report 1992*, Oxford: Oxford University Press.

World Commission on Environment and Development (1987), *Our Common Future*, Oxford: Oxford University Press.

World Conservation Union, United Nations Environment Programme and World Wide Fund for Nature (1991), *Caring for the   environment*, Gland, World Conservation Union, United Nations Environment Programme and World Wide Fund for Nature.

6 Dynamic Regional Development in the EU Periphery: Ireland in the 1990s[1]

JAMES WALSH

Introduction

The 1990s has been a remarkable decade in the history of economic development in Ireland. Unprecedented rates of growth over most of the decade have lead to a rapid convergence in per capita levels of gross national product (GNP) from just under 60 percent of the EU average (when measured at purchasing power parity rates) in the late 1980s to approximately 90 percent in 1998. This dramatic performance has attracted much international attention (Gray, 1997) which has given currency to the label "Celtic Tiger", and also some critical assessment from local analysts (Crowley and MacLaughlin, 1997; Sweeney, 1998).

The distinctiveness of the current phase of very rapid expansion can only be understood by comparing it with the situation that prevailed prior to the early 1990s. Following the adoption of export oriented economic policies and programmes in the late 1950s the Republic of Ireland GDP grew at an average annual rate of about 4 percent over the twenty years up to 1980. This was followed by a phase of deep recession in the early 1980s during which there was negligible economic growth, a decline in public and private investment, a reduction in the number at work, soaring unemployment, a resumption of high levels of net emigration, exceptionally high levels of public indebtedness, and for consumers annual inflation rates in excess of ten percent (Duffy, et. al., 1997).

The situation in the 1990s is very different. Between 1993 and 1997 non-agricultural employment increased by 20 percent. Former emigrants and others are returning in large numbers, frequently equipped

117

with skills and experience that are very much in demand (Barrett and Trace, 1998). The unemployment rate has been reduced from 17 percent in 1987 to 6 percent in 1999. The government current account has been in surplus over recent years and inflation has declined to very low levels. Furthermore, most analysts are agreed that this favourable situation is set to continue for some more years.

This chapter provides an overview of the processes that have contributed to the recent growth and, in particular, it assesses the spatial aspects of the current phase of development.

Underlying Processes

This section attempts to identify the processes underlying the recent transformation of the economy drawing on a number of recent analyses (Bradley *et al*, 1997; Leddin and Walsh, 1997). The indicator that attracts most attention is the per capita GNP. The trend in the index over time is due to the interaction of four inter-related factors: productivity per worker, employment rate, participation rate in the labour force and demographic dependency (Duffy, *et al.*, 1997).

A disaggregated analysis of the trend in per capita GNP reveals that gains in productivity alone do not account for the long-term trend since the 1960s. Throughout most of the past thirty years productivity growth has been strong with average annual growth rates of about three percent. However, this trend did not lead to an improvement in per capita GNP until the 1990s. Participation rates in the labour force were in decline until 1988 after which they have risen sharply. The employment rate was in decline until 1994 after which it increased as unemployment declined. The demographic dependency ratio only started to decline in the mid 1980s.

The distinctive aspect of the 1990s experience is that each of the four factors have made net positive contributions to overall growth in per capita GNP (Fitzgerald *et al.*, 1999). Productivity growth rates of approximately 3.0 percent per annum are associated with the strength of the supply side of the economy driven by the increase in the stock of human capital resulting from long-term educational and training investments and by recent investments in physical infrastructure. These activities have in turn contributed to additional employment. Labour force participation rates have increased mainly due to greater participation of females as Ireland catches up with the more developed parts of Europe (Fahey and Fitzgerald, 1997). Dependency rates have fallen dramatically following the decline in fertility and the recent upsurge in net in-migration (Punch and Finneran, 1999).

Taken together the four factors referred to above help to identify the different sources of per capita growth in output. However, in order to understand how the recent experience came about it is necessary to look for a wider reference frame. Key concerns are how the policy making and delivery systems operate; the priorities in the recovery strategies; the impact of the EU; the dividends from a late demographic transition and from investment in education; and the role of sectoral initiatives. The role of each of these factors will be considered next.

Negotiated Governance

The severity of the crisis in the mid 1980s was detailed in a special report from the National Economic and Social Council (NESC) in 1986. The membership of the Council includes representatives of key government departments, private sector business interests, the agricultural organisations, the trade unions, and government nominees. Following a systematic in-depth analysis of trends over the period from 1980 the Council concluded very pessimistically that

> the economic and social problems now confronting the country are extremely grave. This is particularly the case with regard to the labour market situation and the public finances (NESC, 1986).

In response the NESC proposed a strategy aimed at correcting the imbalances in the public finances, promoting the development of the traded sectors and progressive reform of the taxation rules in order to enhance the efficiency and equity of the system. Given that the burden of adjustment would have to be shared by all sectors of the economy and society it was considered essential that a programme should be initiated to remove many of the inequities impacting on different groups in society. The 1986 NESC analysis was a critically important input to the process that lead to the achievement of a broadly based consensus that facilitated a transition to a new model of governance based on negotiation between the principal stakeholders.

Following the consensus achieved in preparing the NESC 1986 strategy a Programme for National Recovery 1987-1990 was agreed between the government, employers and the trade unions. Three subsequent programmes, covering the periods 1990-93, 1994-97 and 1997-2000, have also been agreed by the partners. A central feature of these agreements has been the inclusion of central wage agreements. These have helped to moderate pay settlements during a period of rapid growth. Agreed procedures that have been adopted in relation to the handling of industrial disputes have resulted in a significant reduction in the number of work days lost to labour disputes. In return reforms have been introduced to lower the burden of taxation and commitments have been entered into in relation to the maintenance of public services. The outcome has been an increase over the decade to 1997 of about 15 percent in real terms in gross industrial earnings which has been supplemented by an additional 9 percent rise in after-tax earnings (Leddin and Walsh, 1997).

The essence of the negotiated governance model, in contrast to the neo-liberal model involving deregulation of markets, is that it relies on a partnership approach which

> provides a procedural and institutional framework for turning the interaction of community, market and state in a constructive direction (O'Donnell, 1997, p.554).

The role of the state is redefined within the partnership approach. Devolution of policy functions requires the state to move from a hierarchical approach in the exercise of its authority and power to a negotiated approach based on networking and co-ordination. It also requires the government to guarantee democratic accountability by maintaining sufficient autonomy to monitor and evaluate the effectiveness of new arrangements. The negotiated governance model that has been adopted at both central and local levels in Ireland has been positively endorsed in a review undertaken for the OECD (Sabel, 1996).

Of course, it is not surprising that the negotiated governance experiment has not been without limitations (NESF, 1997, Walsh, *et al.*, 1998). The following are some of the major concerns that have arisen about the partnership approach: partnerships have not been sufficiently inclusive; there is a real need for capacity building among partners so as to improve their effectiveness; there has been a reluctance to share power and information; there is a sense among some partners that consultation does not lead to any significant shifts in the status quo. Ultimately, negotiated governance via partnerships requires a fundamental change in relations with the emphasis on building trust and sharing power and responsibility. There is also a need for greater mainstreaming of initiatives that have been left to local partnerships. Such a radical shift will only come about through a gradual process involving pilot experiments where the status quo is challenged and potential benefits of new practices can be exemplified.

Macro Economic Stabilisation Policies

In the mid 1980s macro economic policies were required to address to the fiscal and labour market crises. Commencing in 1987 the government introduced major cuts in public expenditure so that current government spending fell by 11 percent in real terms between 1987-89. The growth in the public debt/GNP ratio was reversed and the annual deficit in the balance of payments current account which averaged at almost 8 percent of GNP between 1981-85 was lowered to an average of 0.6 percent for the period 1986-90. Between 1985 and 1988 public sector investment declined by 47 percent in real terms and the real value of public sector grants to enterprises declined by 28 percent. Such drastic cuts, which might have been expected to trigger a deeper recessionary impact, were implemented with the support of the trade unions and the private business interests.

The phase of severe fiscal retrenchment coincided with a marked improvement in trade related to expanding demand in the markets of the main trading partners. This was also helped by a devaluation of the Irish currency in 1986 leading to lower inflation and an improved competitive position in export markets (Leddin and Walsh, 1997).

The main effect of devaluing the currency and curtailing public expenditure in a manner that did not threaten the broad consensus was to restore confidence among investors, especially from overseas. Investment grew at an average annual rate of 3.8 percent between 1985-90 compared with a decline of 3.2 percent over the previous five years. The balance of trade as a percentage of GNP moved into a surplus position for first time since the early 1960s.

The initial impact of the macroeconomic policies was largely confined to overcoming the fiscal crisis. The late 1980s were regarded by some as a period of jobless growth during which total employment increased only marginally and unemployment continued to grow. The severity of the labour market crisis was reflected in a resumption of very high levels of emigration especially to the UK, US, and Australia which were then experiencing a recovery in their economies (Walsh, 1992). However, even though total employment increased by only 0.6 percent between 1986 and 1989 there was in fact a significant restructuring underway between private and public sector employment. Public sector employment declined by 24,000 (7.7 percent) while private sector employment grew by 48,000 (8.2 percent). There was also a reduction of about 11,000 in the number employed in state-sponsored employment schemes.

The macroeconomic framework, established in the late 1980s, has been maintained by successive governments. They have put in place strategies aimed at enhancing international competitiveness and stimulating further growth in the economy. The impact on employment first became evident in 1990 but it was 1994 before the first reduction in unemployment occurred. Long-term unemployment (over three years) only began to decline in 1997 following the introduction of targeted measures specifically for this group.

European Union Impact

The launch of the recovery strategies for Ireland in 1987 coincided with two important milestones in the evolution of the European Union. The adoption of the Single European Act in 1987 committed the Community to the establishment of a single market by January 1993 and also to a reform of the Structural Funds. Both initiatives were of major significance for Ireland (NESC, 1989). This report concluded pessimistically that "the Structural Funds, as currently constituted, will not be sufficient to create convergence.." (p.527). However, an alternative conclusion was reached in an *ex ante* evaluation of the Single European Market by Bradley *et al.,* (1992), which concluded that expanded market opportunities, especially for manufacturing, and increased efficiency in the domestic market particularly for services would lead to an overall increase of five percent in GNP by the year 2000. Subsequent analysis has shown that their forecasts have been exceeded (Fitzgerald, 1998).

The impact of the Structural Funds has been examined at the national level by Duffy *et al* (1997) and Honahan *et al* (1997), while Walsh and Meldon (1995) have considered the regional impacts. Ireland has consistently received the highest level of per capita assistance of all the Objective One regions. The funds have been used to enhance the supply side of the economy by aiding education and training programmes and also by supporting public investments in physical infrastructure. Assistance has also been targeted at overcoming deficiencies in the processing and marketing of output from the productive sectors and towards improving the quality of the product range in areas such as tourism. A significant portion is also targeted at improving the range of skills in the labour force. The expected long run impact of the two rounds of Structural Funds, 1989-1993 and 1994-1999, is estimated to be an increase of about two percentage points in the level of GNP above what it would be without the Funds (Duffy *et al.*, 1997,). Significantly, the impact of Structural Funds in boosting GNP is likely to be much less than the Single Market impact over the longer term.

The reform of the Structural Funds had another significant impact in Ireland which relates to the manner in which the administrative and political systems have had to adapt to requirements set by the European Commission. The partnership model is central to the design and implementation of EU assisted Community Support Frameworks. Eligibility for assistance is contingent on the preparation of plans consisting of integrated multi-annual programmes. All programmes are subject to ongoing monitoring plus *ex ante* and *ex post* evaluations. These requirements have necessitated a very significant change in planning and administrative practices. EU concerns in relation to the natural environment and social policies have been reflected in the most recent Community Support Framework. Effective use has been made of Community Initiatives to pilot test new models that might not have happened without EU support.

Social Dividends from late Demographic Transition and Investment in Education

The demographic transition to low levels of fertility did not commence in Ireland until about 1980, some 15-20 years after many other European countries. One of the consequences of a delayed transition was that productivity gains in the 1960s and 1970s did not translate into increases in per capita output. Female participation rates remained low and high levels of dependency persisted while birth rates were high.

The increase in female participation rates from 29 percent in 1980 to over 40 percent in 1998 has been greatly assisted by improvements in the level of educational provision. Large-scale investment in post-primary and third level education only commenced in Ireland in the late 1960s. By the mid 1990s over 80 percent of those leaving the Irish education system had completed secondary education and 50 percent had experienced third level education. The benefits from such investment have been a substantial contributory factor to the recent economic transformation (Fahey and Fitzgerald, 1997). The main growth areas within the third level educational sector have been in new technologies (electronics and IT), business and finance studies, and modern European languages.

Sectoral Initiatives

The combination of factors that have already been referred to provided ideal conditions for new investments. The major areas of growth in output have been in manufacturing and international services while very significant employment growth has occurred in tourism and also in professional services, retailing, and personal services. Here it is possible to consider only briefly the initiatives that have been undertaken in some of the key sectors.

Manufacturing and Traded Services

A distinction can be made within manufacturing between indigenous enterprises and those that result from inward investment. The indigenous component of Irish manufacturing has been for the most part characterised by small scale, limited technology and heavy reliance on either the domestic or the UK markets. The only major exception to this generalisation has been the dairy processing industry. Following the removal in 1978 of the limited protection given to indigenous manufacturing during the first five years of EEC membership, there were widespread losses in employment among firms in traditional sectors such as footwear, leather goods, textiles, and food and drinks, as the forces of international competition proved to be too great even among firms that tried to become more efficient.

Following a reorientation of industrial policy more towards the indigenous sector in the second half of the 1980s there has been a significant recovery in both employment and output: employment increased by 12 percent between 1987-95, while output grew by 19 percent in real terms between 1988-94 (Breathnach, 1998). What remains of indigenous Irish industry is now much more efficient and competitive (O'Malley, 1998) though much remains to be done in relation to fostering a culture of networking and learning (NESC, 1996, Jacobson and Mottiar, 1999). Niche opportunities afforded by the Single European Market have been targeted successfully. Additionally, there is a growing market for local sub-supply linkages as many foreign controlled firms are encouraged to become more embedded in the Irish economy. The software industry is a prime example of recent dynamic growth which has been assisted by the establishment of Ireland as a major base for international computer industries.

Ireland has a long tradition of attracting inward investment following a major reorientation of industrial policy in the late 1950s. A vast amount of experience has been accumulated in regard to the identification of the factors that influence the location of mobile investments and in evaluating projects. Consequently, the Industrial Development Agency, IDA Ireland, which has responsibility for attracting inward investment, is one of the most sophisticated agencies operating in this field. Following very critical reviews of industrial policy in 1982 and again in 1992, IDA Ireland has adopted a more selective sectoral approach. It has placed its main emphasis on technically sophisticated activities such as electronics, pharmaceutical and healthcare sectors, and also on international services which are reliant on excellent telecommunications and skilled graduates in modern languages. In addition to focusing on a limited number of sectors the agency has also targeted key international players in these sectors (Breathnach, 1998). Most analysts agree that inward investment has been the main driver of the exceptional economic performance in the 1990s (see Murphy, 1997; or Breathnach, 1998, for a more detailed explanation of the current surge in inward investment).

In 1998 there were 1140 overseas companies operating in Ireland with IDA support compared with 670 in 1987. Total full-time employment in these companies has reached almost 116,000, up almost 80 percent since 1987. Between 1994 and 1998 there was an increase of almost 35,300 (53 percent) in the total number employed in the pharmaceuticals/healthcare, electronics/engineering and international services sectors with the latter two sectors accounting for 89 percent of the total increase, in almost equal proportions.

A major breakthrough was achieved in 1989 with the attraction of Intel, the world's leading producer of microprocessors which has set up its European manufacturing base at Leixlip, west of Dublin. This had a significant demonstration effect encouraging many other leading firms to establish operations in Ireland.

In addition to inward investment in manufacturing there has also been considerable growth in the internationally traded services. The principal drivers in the growth of services have been the availability of a high quality telecommunications infrastructure, competitive rates for international telephone business communications, and availability of a labour force proficient in European languages. Total employment in the sector increased from 3,600 in 1987 to over 25,500 in 1998. An International Financial Services Centre was established in Dublin in 1987 to act as a base for international banking, insurance, leasing and fund management. Foreign owned software operations have grown rapidly, led by Microsoft, Lotus and Oracle so that Ireland is now the second largest exporter of software after the US. Furthermore, since the early 1990s the

IDA has targeted international teleservices that require a labour force proficient in technical and continental language skills. This is regarded as one of the fastest growing sectors with an employment projection of 10,000 by the year 2,000 compared with 3,400 in 1996. A special training programme for school leavers targeted at this sector has recently been launched in order to overcome a skills shortage.

Other Services

The rapid growth in the productive sectors has fuelled an enormous increase in the demand for professional services in both the private and public sectors. Furthermore, increases in disposable incomes and growing numbers at work have lead to significant growth in demand for retail and personal services. Additionally, there has been a tourism boom with growing numbers attracted from new destinations such as mainland European countries. The market has also been augmented by growth in business travel and the availability of cheap flights to Dublin for short vacations. Key attractions for the new tourists are Ireland's culture, historic heritage and a largely unspoilt rural environment. The quality of much of the ancillary tourism product has been upgraded with the assistance of EU Structural Funds. The contribution of tourism to GNP is approximately 8 percent. As a labour intensive industry it has made a significant contribution to the growth in employment, especially in some rural regions (Hannigan, 1997).

The Spatial Dimensions of Economic Change

This section examines the geography of the current phase of economic development. For the most part the discussion is confined to NUTS III regional level changes due to the unavailability smaller scale data. Furthermore, only a small number of indicators can be considered here, (for a more detailed analysis see the contribution by the present author in Fitzgerald, *et al.,* 1999, Boyle, McCarthy and Walsh, 1999).

Output Indicators

In 1996 the first set of regional accounts was published by the Central Statistics Office (CSO, 1996). Estimates of Gross Value Added (GVA) have been made for each of eight NUTS III level regions, corresponding with the Regional Authority Areas (see Walsh, 1995), where GVA is defined as a measure of the value of the goods and services produced in the region at the value which the producer received minus any taxes payable and plus any subsidies received as a consequence of their production or sale. Some caution is required in the interpretation of the data as the GVA estimate includes the operating profits of overseas companies, even though a large portion of these may not be reinvested in the region. The impact on the regional estimates is to increase the disparities between the greater Dublin region, and also the Southwest (which includes a concentration of electronics and pharmaceutical plants around Cork city) on the one hand, and the Midland, Border and West regions on the other hand. The GVA definition given above is calculated on basic prices, it omits subsidies such as headage payments, livestock premia, payments to farmers under the Rural Environment Protection Scheme. When the latter are included the differences between the more urbanised and the predominantly rural regions are somewhat smaller (Boyle, *et al.*, 1999).

The regional estimates show significant variations in per capita GVA and also that there has been a tendency towards divergence (Table 6.1). The 1995 per capita index for the Dublin and the Mid-East (it is essential to calculate a combined index for the two region in order to overcome the effects of commuting from the Mid-East to Dublin) was 21 percent above the average for all regions whereas the indices for the west and Midland regions were only 70 and 72 percent of the overall average. The Border region index was only marginally higher due in large part to the presence of a small number of overseas companies that are generating very substantial amounts of value added.

Table 6.1 Regional Levels of Gross Value Added 1995 and Change 1991-1995

Region	Per capita GVA 1995 (£)	Percentage change 1991-95	Index 1991	Index 1995
Dublin & Mid-East	12,005	42.4	118	121
Southwest	10,530	40.1	105	106
Midwest	9,360	39.3	94	94
Southeast	8,539	32.4	90	86
Border	7,658	29.6	82	77
Midlands	7,145	30.9	76	72
West	6,950	29.5	75	70
Ireland	9,917	38.4	100	100

Calculated from CSO Regional Accounts data

The differences between regions can be decomposed into two sets of factors: productivity levels and activity ratios. The latter measure the proportion of the population in each region that are at work. Activity ratios reflect differences in employment rates, labour force participation rates and dependency levels. The total effect of differences in activity ratios on the distribution of per capita GVA is small amounting to about one-seventh of the gap between the richest and poorest regions.

The main source of variation is in productivity levels. There are differences between regions in the sectoral distribution of employment and also in sectoral productivity levels. Primary activities account for 17 percent of total employment in the three poorest regions. The service sector share of employment is lowest in these regions at 50 percent overall. Manufacturing and construction account for just over 30 of total employment in the Southeast, Southwest and Midwest regions compared with 25 percent in the East (Dublin and Mid-East) where services are the dominant source of employment at 73 percent.

The contribution of each sector to total GVA differs from the distribution of employment across sectors reflecting sectoral differences in productivity. Primary activities accounted for only 5.8 percent of GVA in 1995 compared with 11.5 percent of employment. For manufacturing and construction the corresponding proportions were 42 percent and 28 percent. Taking the average GVA for all persons at work in 1995 as 100, the sectoral indices are 51 for primary activities, 150 for industry and 86 for services (Table 6.2). In the West region the level of GVA per worker in the primary sector is only 27 percent of the overall average. By contrast, above average productivity levels in this sector are found in the more commercial dairy and tillage farming regions in the Southwest, Southeast and Mid-West.

Table 6.2 Indices of Gross Value Added per person employed by sector and by region 1995

Region	Agriculture*	Industry	Services	Total
Dublin & Mid-East	49	177	98	116
Southwest	65	195	77	110
Midwest	55	126	81	92
Southeast	55	142	72	89
Border	59	105	72	81
Midland	46	80	80	74
West	27	111	73	71
Ireland	51	150	86	100

* includes forestry and fishing
Calculated from CSO Regional Accounts data

In the industrial sector where overall productivity levels are highest the Southwest and East regions have significantly higher levels mainly due to the concentrations of high value-added manufacturing sectors around Dublin and Cork city. By contrast, productivity levels in manufacturing in the West, Border and Midlands are particularly low.

Services account for a little over half (52 percent) of total GVA for all regions. There is a significant difference between the East and all other regions in this sector reflecting the level of concentration of high value added producer services in Dublin. GVA levels per worker in services are particularly low in the Border, Southeast and West regions - in the latter two regions there is a high level of dependence on tourism.

Between 1991 and 1995 total GVA is estimated to have increased by 41 percent with 52 percent of the increase occurring in industry and 46.5 percent in services. Total per capita GVA increased by 38.3 percent over the same period with approximately 76 percent of the gain attributable to improved productivity.

Indexing the rate of increase for all regions to 100 the East has by far the highest with an index of 121 followed by the Southwest (104) and the Mid-West (103). The West and Border regions each have indices of 77 while the Midlands (80) and Southeast (84) are only marginally better. The East benefited from a further improvement in its already very high productivity levels and a continuing high activity ratio. In the Southwest a slight decline in the activity ratio was more than offset by a further improvement in productivity. The opposite occurred in the Mid-West. Relative total productivity indices declined in each of the other regions with only very minor compensations related to improved activity ratios in the Midlands and Border regions.

Household Incomes

Regional estimates of per capita GVA are output measures and should not be regarded as proxies for living standards. Data from the *Household Budget Surveys* undertaken by the Central Statistics Office show that average weekly household disposable incomes (AWHDI) increased in real terms by 12.7 percent between 1987 and 1994/5. However, the pattern of change was very uneven (Table 6.3). By far the largest increase was recorded for the Midland region, which had the second lowest level in 1987. Broadly similar increases of around 17-18 percent occurred in the West, Border and East regions. On average households in the Mid-West, Southwest and Southeast benefited from only marginal increases. These data suggest that outside the East region some convergence has taken place among households in the remaining regions. The overall position in 1995 was that the AWHDI index in the East was 115 (Ireland = 100) followed by the Mid-East, Midlands and West (all between 97 and 106) with the lowest in the Southeast (88), Border (90) and Mid-West (90). The HBS also confirms that there is a substantial amount of regional redistribution of income taking place through the taxation system and the range of publicly funded welfare programmes and other state transfers (Boyle, et al., 1999).

Table 6.3 Indices of average weekly household disposable income (AWHDI)

Region	AWHDI (£) 1994/5	AWHDI Index 1987	AWHDI Index 1994/5	% change in AWHDI 1987-1994/5*
Dublin	325.14	111	115	16.9
Mid-East	297.84	101	106	18.2
Midlands	273.93	86	97	27.8
West	273.43	92	97	18.4
Southwest	264.61	101	94	5.0
Midwest	255.45	101	91	1.2
Border	252.92	87	90	16.9
Southeast	247.49	94	88	5.0
Ireland	281.92	100	100	12.7

* measured in real terms to take account of inflation
Data source: Household Budget Surveys 1987 and 1994/5 CSO. Dublin

Data on incomes for farm households reveal a wide variation and an increasing tendency to rely on direct payments (Frawley, 1998). This is especially the case throughout most of the West and Border regions where low intensity cattle and sheep farming systems are prevalent (Lafferty, Commins and Walsh, 1999). The prospects for agriculture in Ireland under the Berlin agreement on CAP reform are not encouraging except perhaps for the larger dairy farms in the Southwest (Donnellan, *et al.*, 1999). The share of agricultural income derived from direct payments is set to increase 70 percent - this prospect coupled with the increased demand for labour in other sectors of the economy will lead to further decline in the number of younger farmers. In this context there is a strong need for an integrated rural development strategy linked to a broader framework for regional development (NESC, 1995; Government of Ireland, 1997a).

Despite the general improvement in the economy there are many living on very low incomes. The central wage agreements that have been in place since 1987 have not been able to bring about a reduction in wage disparities. Wage dispersion has actually increased so that Ireland has the most unequal earnings distribution among 16 OECD countries. The ratio of earnings of the top decile to the bottom for men working full-time increased from 3.5 to 5.0 between 1987 and 1994 (Nolan and Hughes, 1997).

The disparities in wage levels can be related to increasing social polarisation within the labour force. A distinction can be made between skilled, well educated professional workers who are in core areas of employment and those who are on the periphery in poorly paid forms of atypical work, e.g., part-time or temporary workers, particularly females in tourism, retailing and personal services.

Despite unprecedented levels of economic growth unemployment has remained at high levels and poverty is still a very real issue. The highest risk of poverty was observed for villages and towns with populations of less than three thousand (Nolan *et al*, 1998). Problems of social exclusion, poverty and disadvantage have been highlighted in many reports instigated by the Combat Poverty Agency (e.g., Curtin *et al.*, 1996; Pringle *et al*, 1999; Nolan and Watson, 1999). On a more general level Clark and Kavanagh (1996) have drawn attention to the inadequacy of economic indicators as a measure of social progress.

Employment Patterns

The regional distribution of employment gains and losses is an important issue in debates on regional development. While for many years the main concern of some commentators was the perceived phenomenon of jobless growth, the major issue since about 1997 has been the spatial distribution of the additional employment associated with economic expansion.

According to the 1997 Labour Force Survey (LFS) there were 1.338 million persons at work, almost one-quarter of a million more than in 1988 (CSO, 1998). Most of the increase (78 percent) has occurred since 1993. Between 1993 and 1997 total employment increased by 186,200 of which 52 percent occurred in the East and only 20 percent in the three weakest regions. The corresponding percentages for the period 1989-93 were 38 and 20 respectively.

The trend in non-agricultural employment is more useful as a guide to the prospects for each region (Table 6.4). Total non-agricultural employment increased by 82,400 between 1989-93 and a further 196,100 between 1993-97. However, a marked divergence has occurred between the two sub-periods. Between 1989-93 approximately 30 percent of the total increase took place in the East while there was an almost equivalent gain in the West, Border and Midland regions combined. By contrast, in the later period almost half of the total increase occurred in the East while only 22 percent took place in the weaker regions. The proportion for these regions would be even lower were it not for the emergence of long distance commuting from many parts of the Midland region to Dublin. One of the impacts of such rapid growth in employment in and around Dublin, coupled with an abundant supply of mortgage finance at low interest rates, has been a very significant surge in the demand for, and the cost of, housing (Bacon and Associates, 1998).

Table 6.4 Change in non-agricultural employment 1989-1997

Region	Actual change 1989-93	Actual change 1993-97	Percentage change 1989-93	Percentage change 1989-93
Dublin	16,500	61,800	4.88	17.43
Mid-East	8,200	35,300	9.81	38.45
Southwest	13,900	29,300	10.81	16.09
Midwest	8,800	13,000	21.14	15.80
Southeast	11,300	13,000	12.70	12.96
Border	9,500	15,300	10.24	14.96
Midlands	6,000	14,900	14.02	29.92
West	8,200	13,500	10.83	16.09
Ireland	82,400	196,100	8.91	19.47

Calculated from Labour Force Surveys, 1989, 1993, 1997. CSO. Dublin

Recent data from Forfas - the policy advisory and co-ordination board for industrial development and science and technology in Ireland - on employment in State-assisted companies shows that 70 percent of the total net gain in employment in these companies between 1993-97 came from overseas companies. On a regional level the data show that 69 percent of the total net gain took place in the East - the proportion is the same for both Irish-owned and overseas companies. Another 23 percent of the net gain in employment in overseas companies occurred in Cork, Limerick and Galway. The data also highlight the weaknesses of the Irish owned sector in the West and Border regions. Significant net losses among these companies almost totally offset the gains among overseas companies so that the net increase in employment for all State assisted companies in the two regions amounted to only 3.4 percent of the total increase for all regions.

Demography and Settlement

Ireland has been undergoing a delayed demographic transition. This has been associated with a significant decline in fertility since 1980. Trends in net migration have fluctuated with significant levels of in-migration being experienced at present in contrast to very high levels of out-migration in the late 1980s. Many are highly skilled and responding to opportunities in the labour market (Barrett and Trace, 1998) especially in Dublin. There has also been an increase in participation in higher education which has been accompanied by higher levels of inter-regional mobility. Many aspects of the recent changes in the distribution of population that are relevant to regional and local development are discussed in Walsh (1996).

The city regions are able to provide large numbers of skilled workers (many of whom have migrated from other parts of the state, McHugh and Walsh, 1995), serviced business parks, high international accessibility, contacts with universities and research centres, a wide range of support services and attractive and diverse social and physical environments. As Ireland has a very weak urban system - only four centres have more than 50,000 inhabitants - there are particularly severe problems of adjustment in many rural areas that are too distant from the main centres for daily commuting (Walsh, 1997). More details on the problems related to the settlement pattern in Ireland are provided in NESC (1997) and Walsh, (1998).

The Environment

The pace and uneven regional distribution of economic expansion has placed considerable pressures on the environment. Difficulties have arisen in relation to increased use of energy from non-renewable resources and related atmospheric pollution, traffic congestion especially in Dublin, rezoning of lands for residential purposes on the fringe of Dublin, tourism projects especially in ecologically sensitive areas, infrastructure investments such as roads, waste management sites and sewerage treatment facilities, and new rural land uses such as forestry. Efforts to maintain household incomes in some rural areas through direct payments have led to overgrazing by sheep and destruction of local habitats. Large scale tourism-related projects in urban areas have impacted on the architectural integrity of streetscapes while commercial property developments have turned some of the prime retailing areas into replicas of what can be found in most international cities. In this way there has been some loss of local identity and distinctiveness. Acrimonious debates have occurred as some have sought to trade off the environment against investment in projects that would allegedly lead to additional employment. While there is an official commitment to sustainable development as exemplified by the *National Development Plan 1994-99* and *Sustainable*

Development: A Strategy for Ireland (Government of Ireland 1997b) the reality is that the full implications of a sustainability strategy are poorly understood.

Discussion and Conclusions

The experience of dynamic economic adjustment in Ireland over the past decade shows that a weak underdeveloped and very open region in the European periphery can undergo a rapid transformation that will reduce the economic differentials between it and the core regions of the EU. The transformation cannot be explained by any single factor - rather several diverse factors have interacted. A model of governance that facilitated collaboration and partnership between the state, the business sector and representatives of other interests in society was essential. The impact of EU integration has on the whole been benign. Additional opportunities provided by the Single Market and the inducement to eliminate inefficiencies have more than offset the losses in uncompetitive firms, while the Structural Funds provided the resources at the correct time to enhance the quality of supply-side supports and upgrade much of the physical infrastructure. Along with the aforementioned factors strategic positioning and repositioning vis-a-vis globalisation processes was a very significant factor in the attraction of inward investment which has been the main driving force behind the growth in output. The location pattern of new inward investment in key sectors of manufacturing and internationally traded services has influenced many other public and private decisions which have contributed to the emergence of new regional patterns with in most cases significant intra-regional contrasts. The critical nature of the relationships between the environment, economic development and social equity have been brought more sharply into focus, presenting major challenges for planning in all spheres in the years ahead.

The experience over the past decade has also demonstrated clearly the problems that can arise in the absence of a strategy for spatial development. This was especially significant given the reliance on inward investment which tended to concentrate around Dublin and to a lesser extent around Cork, Limerick and Galway cities. Serious problems have arisen in relation to congestion, availability of housing and skills shortages as well as in the planning process for infrastructural investment. There is now an urgent need for a strategy that will provide a framework for balanced regional development while at the same time maintaining the competitiveness of firms within the international economy. The division of

Ireland into two NUTS II regions (one with Objective One status) for the period 2000-2006 will ensure a greater allocation of investment for infrastructure and the productive sectors in the weaker region. However, the distribution of investments will need to be guided by a prioritised strategy for spatial development. The minimum requirements for a spatial development strategy have been set out in Fitzgerald *et al.*, (1999) and more specific proposals in relation to settlement and infrastructure are contained in two reports prepared by Fitzpatrick Associates (1999a,b). These will act as an input to the preparation of the *National Development Plan* for the period 2000-2006. In addition Strategic Planning Guidelines for the Greater Dublin Area have been prepared (Brady, Shipman, Martin Consultants, 1999). Unless there is a rapid official response to these reports the prospect is for increasing economic convergence between Ireland and the EU core regions but also for a further widening of the regional prosperity gaps within Ireland.

Note

1   Some of the material included in this chapter was already published as part of an article by the author in *Geographical Viewpoint*, vol. 25, 1997.

References

Bacon, P. and Associates, (1998), *An Economic Assessment of Recent House Price Developments*. Dublin: The Stationery Office.

Barrett, A. and Trace, F., (1998), Who is coming back? The educational profile of returning migrants in the 1990s. *The Irish Banking Review*, Summer '98, 38-52.

Boyle, G. E., McCarthy, T. and Walsh, J. A., (1999), Regional income differentials and the issue of regional equalisation in Ireland, *Journal of the Statistical and Social Inquiry Society of Ireland*, vol. XXVIII, Part 1. (in press).

Bradley, J. et al., (1992), *The Role of the Structural Funds: Analysis of Consequences for Ireland in the Context of 1992*. Policy Research Series No. 13. Dublin: The Economic and Social Research Institute.

Bradley, J., Fitzgerald, J., Honahan, P. and Kearney, I., (1997), Interpreting the recent Irish Growth Experience. In Duffy, D., Fitzgerald, J., Kearney, I., and Shortall, F., *Medium Term Review 1997-2003*. Dublin: The Economic and Social Research Institute.

Brady, Shipman, Martin Planning Consultants, (1999), *Strategic Planning Guidelines for the Greater Dublin Area*, Dublin.

Breathnach, P., (1998), Exploring the 'Celtic Tiger' Phenomenon: causes and consequences of Ireland's economic miracle, *European Urban and Regional Studies*, 5,4, 305-316.

Central Statistics Office, (1996), *Regional Accounts 1991*. Dublin: The Stationery Office.

Central Statistics Office, (1998), *Labour Force Survey 1997.* Dublin: The Stationery Office.

Central Statistics Office, (1998), *Census 96 Principal Socio-economic Results.* Dublin: The Stationery Office.

Clark, C. and Kavanagh, C., (1996), Progress, Values and Economic Indicators. In B. Reynolds and S. Healy (eds.) *Progress, Values and Public Policy.* Dublin: CORI.

Crowley, E. and MacLaughlin, J. (eds.) *Under the Belly of the Tiger: Class, Race, Identity and Culture in the Global Ireland,* Dublin: Irish Reporter Publications.

Curtin, C., Haase, T. and Tovey, H., (1996), *Poverty in Rural Ireland,* Dublin: Oak Tree Press.

Donnellan, T., Binfield, J. and McQuinn, K., (1999), *Impact of the Berlin Agreement on Irish Agriculture.* Dublin: Teagasc.

Duffy, D., Fitzgerald, J., Kearney, I. and Shortall, F. (1997), *Medium Term Review: 1997-2003,* Dublin: The Economic and Social Research Institute.

Fahey, T. and Fitzgerald, J. (1997), *The Welfare Implications of Demographic Change.* Dublin: Oak Tree Press.

Fitzgerald, J. 1998. An Irish Perspective on the Structural Funds, *European Planning Studies,* 6,6, 677-694.

Fitzgerald, J., Kearney, I., Morgenroth, E. and Smyth, D. (1999), National Investment Priorities for the Period 2000 - 2006, Dublin: The Economic and Social Research Institute.

Fitzpatrick Associates, (1999),a. *Border, Midland and Western Region Development Strategy 2000 - 2006.* Dublin.

Fitzpatrick Associates, (1999),b. *Southern and Eastern Region Development Strategy 2000 - 2006.* Dublin

Frawley, J., (1998), *The Impact of Direct Payments at Farm Level: a county study.* Dublin: Teagasc.

Government of Ireland, (1997),a. *Report of the Rural Development Policy Advisory Group.* Dublin: The Stationery Office.

Government of Ireland, (1997),b. *Sustainable Development: A Strategy for Ireland.* Dublin: The Stationery Office.

Gray, A. W. (ed.) (1997), *International Perspectives on the Irish Economy.* Dublin: Indecon Economic Consultants.

Hannigan, K. (1997). Tourism Policy and Regional Development in Ireland. In McCafferty, D. and Walsh, J. (eds.) *Competitiveness, Innovation and Regional Development in Ireland.* Dublin: Regional Studies Association (Irish Branch), 171-192.

Honahan, P. (ed.). (1997), *EU Structural Funds in Ireland: a mid-term evaluation of the CSF 1994-99.* Dublin: The Economic and Social Research Institute.

Jacobson, D. and Mottiar, Z., (1999), Globalization and modes of interaction in two sub-sectors in Ireland, *European Planning Studies,* 7,4, 429- 444.

Lafferty, S., Commins, P. and Walsh, J., (1999), *Irish Agriculture in Transition: a Census Atlas of Agriculture in the Republic of Ireland.* Dublin: Teagasc.

Leddin, A. and Walsh, B., (1997), Economic stabilisation, recovery and growth: Ireland 1979-1996. *The Irish Banking Review,* Summer '97, 2-18.

McHugh, C. and Walsh, J.A. (1995), The Irish School leaver trail: links between education and migration. *Geographical Viewpoint,* 23, 88-103.

Murphy, A., (1997), *The Celtic Tiger - the Great Misnomer.* Dublin: MMI Stockbrokers.

National Economic and Social Council, (1986), *A Strategy for Development,* Dublin: The Stationery Office.

National Economic and Social Council, (1989), *Ireland in the European Community: Performance, Prospects, Strategy,* Dublin: The Stationery Office.

National Economic and Social Council, (1995), *New Approaches to Rural Development,* Dublin: The Stationery Office.

National Economic and Social Council, (1996), *Networking for Competitive Advantage,* Dublin: The Stationery Office.

National Economic and Social Council, (1997), *Population Distribution and Economic Development: Trends and Policy Implications* , Dublin: The Stationery Office.

National Economic and Social Forum, (1997), *A Framework for Partnership - Enriching Strategic Consensus through Participation.* Dublin: The Stationery Office.

Nolan, B., and Hughes, G., (1997), *Low pay: the earnings distribution and poverty in Ireland, 1987-1994,* Dublin: The Economic and Social Research Institute.

Nolan, B., Whelan, C.T., and Williams, J., (1998), *Where are Poor Households? The Spatial Distribution of Poverty and Deprivation in Ireland.* Dublin: Oak Tree Press/Combat Poverty Agency.

Nolan, B. and Watson, D., (1999), *Women and Poverty in Ireland.* Dublin: Oak Tree Press/Combat Poverty Agency.

O'Donnell, R., (1997), Irish policy in a global context: from state autonomy to social partnership, *European Planning Studies,* 5, 4, 545-558.

O'Malley, E., (1998), Revival of Irish indigenous industry 1987-1997, *Quarterly Economic Commentary,* Special Article April, Dublin: The Economic and Social Research Institute.

Pringle, D.G., Walsh, J. and Hennessy, M., (1999), *Poor People, Poor Places: a Geography of Poverty and Deprivation in Ireland.* Dublin: Oak Tree Press/Geographical Society of Ireland.

Punch, A. and Finneran, C., (1999), The demographic and socio-economic characteristics of migrants 1986-1996, *Journal of the Statistical and Social Inquiry Society of Ireland,* XXVIII, Part 1 (in press).

Sabel, C., (1996), *Ireland: Local Partnerships and Social Innovation,* Paris: OECD.

Sweeney, P. (1998), *The Celtic Tiger; Ireland's Economic Miracle Explained.* Dublin: Oak Tree Press.

Walsh, J. A. (1992), Economic restructuring and labour migration in the European Union: the case of the Republic of Ireland. In M. O'Cinneide, and S. Grimes, (eds.) *Planning and Development of Marginal Areas.* Galway: Centre for Development Studies, 23-36.

Walsh, J. A., (1995), *Regions in Ireland: a statistical profile,* Dublin: Regional Studies Association.

Walsh, J. A., (1996), Population change in the 1990s: preliminary evidence from the 1996 Census of Population in the Republic of Ireland, *Geographical Viewpoint,* 24, 3-12.

Walsh, J. A. (1997), Regional Development Challenges. In McCafferty, D. and Walsh, J. (eds.) *Competitiveness, Innovation and Regional Development in Ireland.* Dublin: Regional Studies Association (Irish Branch), 113-131.

Walsh, J. A., (1998), Towards a settlement strategy for sustainable development. In Meldon, J. (ed.) *Learning Sustainability by Doing - Regional Integration by the Social Partners.* Dublin: An Taisce and URGE, 62-64.

Walsh, J. A. and Meldon, J., (1995), Structural Funds and Sustainable Regional Development. In Ireland. In Byron, R., (ed.) *Economic Futures on the North Atlantic Margin,* Aldershot (Hants.) : Avebury Press, 315-27.

Walsh, J., Craig, S., and McCafferty, D. (1998), *Local Partnerships for Social Exclusion?* Dublin: Oak Tree Press.

PART III:

SPATIAL PLANNING, ENVIRONMENTAL MANAGEMENT AND REGIONAL DEVELOPMENT

7 Strategic European Spatial Planning - Power, Knowledge and Rationality in Policy Evaluation

GORDON DABINETT AND TIM RICHARDSON

Introduction

This chapter[1] presents a discussion about an emerging area of spatial policy discourse, European-wide strategic planning. It examines the current policy framework at the European-wide level by exploring the alternative planning paradigms that underpin the case for rationality within this framework. By using such a theoretical review, the paper seeks to argue that the current deployment of evaluation within strategic spatial policy and planning is poorly understood, and further work is needed to develop an understanding of the role of power and knowledge in rationality, and in particular the role they play within the pluralist models of evaluation that are emerging within the wider EU policy processes. This is illustrated by specific evaluation experiences within the EU Trans-European Transport Networks and Structural Funds.

It would appear that we are entering a new era in spatial planning. Momentum appears to be increasing to create a spatial approach at the European level (CEC, 1997a,b), co-existing at present alongside other EU-wide spatial policies (Bachtler & Turok, 1997). New and recent initiatives are providing the first building blocks in a process which, though still hazy and ill-defined, promises to set new challenges for the spatial planning systems in the different EU member states, and the current attempts at trans-national planning, such as INTERREG (eg Faludi, 1997; Hull, 1996; Williams, 1996). However, in this paper, we consider just one element -

the use of evaluation within the EU policy processes - and explore some of the broad conceptual issues raised in developing evaluation in any European-wide spatial planning perspective.

First, the chapter explores how the emerging field of European-wide spatial planning is a contested policy field. We then seek to illustrate how, in the fields of spatial planning, alternative paradigms are contesting the basis of rationality of the policy process and how these might provide a framework to understand more fully the evaluation processes. We then use illustrations from established EU spatial policies articulated through the Trans-European Networks (TENs) and Structural Funds to advance our arguments. At the heart of our exploration is a concern that the deployment of evaluation in spatial planning is shaped by poorly understood power struggles which impose interest-based hegemonies of knowledge. In this way, we postulate that power shapes knowledge and determines policy under the banner of rationality.

Emerging European-wide Spatial Planning

The emergence of EU spatial planning raises important issues concerning its form and legitimacy. In examining future land use regulatory frameworks, Morphet (1997) suggests that Europe has witnessed a renewed interest in strategic spatial planning during the 1990s. An interest which has led to a review of the tools and approaches which might constitute these frameworks, and a convergence of interest in a common European spatial approach. For many years, the EU has implemented measures in a range of sectors which have such spatial effects. A wide range of policies, regulations and other instruments are being used to pursue the key EU objectives of the creation of the single market, the promotion of balanced and sustainable economic and social development, and reductions in disparities between levels of development of different regions. Many programmes respond to problems in particular parts of the EU, such as regions in industrial decline, structurally backward regions, and regions in need of rural and agricultural development. So spatial redistribution is at the heart of many programmes. Alongside these policies with clear spatial aims, other sectoral policies, for example for transport and the environment, have impacts on land use and development. However, the spatial impacts of many of these policies and programmes have often been overlooked in their implementation and evaluation (Davies, 1994).

Furthermore, tensions potentially exist between EU policy objectives and those of national planning systems. The existence of many national systems of land ownership, planning control and building regulations, working with separate and potentially exclusive objectives, could be argued as counter to the Single Market, which depends on member states and other interests pursuing policies and actions which are in harmony with the overall EU integration project. This certainly creates problems in transnational planning situations, and questions the possibility of constructing a planning framework within which these conflicting objectives can be pursued equitably (Williams, 1996). So, the momentum towards EU spatial planning can be seen to linked with the need to deal with not only spatial problems, but also conflicting spatial objectives at different levels and in different regions.

In the 1990s the EU, through Commission initiatives and the work of the inter-Governmental Committee of Spatial Development (CSD), progressed a series of initiatives on spatial co-operation in Europe (CEC, 1997a; Shaw et al, 1995; Williams, 1996). The key developments have been the explorations of European spatial trends and concepts, in Europe 2000 (CEC, 1991) and Europe 2000+ (CEC, 1994), and the Compendium of studies of planning systems in the member states. This work has been buttressed by increasing transnational actions by member states, for example in co-operation over international infrastructure links, and by the increasing political support for transnational planning in the EU institutions, particularly the Committee of the Regions (Williams, 1996).

In 1997, the publication of the First Official Draft of the European Spatial Development Perspective (ESDP), produced by the CSD, marked the next major step towards a coherent approach to EU spatial planning (CEC, 1997b). Together with its further iteration in 1998 (CEC, 1998), the draft ESDP has been the vehicle for wide ranging consultation at the EU level and within member states on the future direction of spatial planning in Europe. It attempts to provide a framework for simplifying co-ordination of transnational planning between member states, but also anticipates in the longer term a broader EU level of planning activity. It intends to encompass urban systems, infrastructure and knowledge, natural and cultural heritage and territorial perspectives, synthesised at EU level. It is also intended that the tools for implementation will form a cascade from the EU to the local level. The ESDP sets the context for mega-regional perspectives and visions to be developed under INTERREG IIc, which constitutes a key step towards a European level of planning, in that it enables cross-border transnational planning initiatives between national and European levels. The ESDP gives a view on the potential for the further institutionalisation of European spatial planning.

The spatial analysis within the ESDP aims to construct a solid patrimony of analyses and interpretation referred to the whole of the European territory (CSD, 1996). The analysis focuses on components of European territory in the three fields of action - (1) urban systems, infrastructures, and natural and cultural heritage, (2) on the spatial impacts of major socio-economic phenomena (e.g. the information society), and (3) on the spatial impacts of sectoral policies. The prospect here is of a dramatic increase in the need for gathering and management of spatial information. For example, member states are likely to be required to contribute analyses of the territorial effects produced at national level by both Community and national policies (CSD, 1996). The definition of co-ordinated and compatible spatial options to be pursued at national, transnational and EU level forms the second core area of work for the ESDP. The options are identified by member states, and are building towards what is again termed a common patrimony (CSD, 1996) where broad support is found for particular options. Compatibility here is defined according to the objectives set within the ESDP, which reflect once again the major EU agendas, and the existing sectoral policies. For example, one key ESDP objective, which closely follows EU policy, is to optimise the contribution of the trans-European networks to economic and social cohesion and sustainable development.

Thus, EU spatial planning is being developed in a complex institutional framework, and will be shaped by major tensions and power relations. It is important then that we do not try to understand this emergence of an EU spatial planning framework as a purely comprehensive scientific rational process, or the benign convergence of national planning systems. It bends to an agenda, and to forces, which contest the future path of development of Europe, and so is likely to have at its core the currently hegemonic ideologies of the single market and political integration, but will also reflect other debates about cohesion and environment (Faludi, 1997; Hague, 1996). Thus the process of creating a European spatial planning framework is not reducible to a technical exercise. It is implicitly normative and ideological - about politics and power as much as about rational policy making. As Giannakourou (1996) claims -

> If the needs of the European integration process seem to have added a European level of spatial planning policy to that of the national states, it is the economic and institutional properties and dilemmas of this same integration process that circumscribe the conceptual identity and the

normative value of the emerging policy... the central question becomes what the conceptual and ideological identity of a European spatial planning policy can be under a market-oriented integration system (p602).

Clearly then, the integration of the objectives and normative positions of the many different levels of government, and other actors, will be an important factor in the shaping of the European spatial approach. It seems important, then, that we should attempt to understand the effects of power relations and normative agendas on the emerging planning framework. We posit that evaluation is a critical element in any such planning framework. We point to the conceptualisation of evaluation, in at least some planning/policy paradigms, as a political tool which responds to the norms and values of the hegemony within which it is deployed. We suggest that this awareness might be used to design evaluations which deliberately seek to avoid this normative loading, and adopt a bottom up approach to the identification and assessment of key values. The key tasks for European spatial planning will be to attempt to resolve or mitigate these critical tensions.

Rationality and Evaluation in Spatial Policy and Planning

Town planning within Britain and other EU member states has emerged out of particular cultural, political and economic circumstances (Greed, 1996; Williams, 1996). Whilst acknowledging our anglo-centric position, it is not the purpose of this chapter to review these constructs, but rather examine spatial planning and processes of land-use policy within a wider interpretation of public policy (Evans, 1997). It is possible to regard the function of a more broadly drawn urban and regional planning as a means of balancing and arbitrating between competing interests in land or space, where (Department of the Environment, 1992):

- land is both a factor of production and an environmental resource;
- the demand for and use of land and territory is a reflection of complex social and economic processes involving a multiplicity of private interests;
- planning draws its objectives from a wide variety of interests, but seeks to strike a balance in the 'public interest';
- in comparison with most other areas, planning involves long timescales and substantial time lags;

- planning operates across a number of scales of government often involving the need to balance local and wider interests; and,
- planning seeks to achieve its aims through a variety of instruments which operate in parallel with other strands of public policy.

Thus planning is never concerned purely with physical land use but must also relate to and allow for other policy dimensions. In this context the spatial plan is seen as a framework in which a set of relationships prioritise certain criteria and action over others and provide a point of reference for those making subsequent decisions (Healey, 1994).

In theory and practice, urban and regional planning has been strongly associated with rationality, and evaluation has been viewed as an essential element within this. Evaluation serves to reconcile the interests which become affected by the planning framework, as decisions change the spatial patterns of economic, social and environmental costs and benefits within an area. Attempts to establish a theoretically grounded view of what evaluation should entail reveals that there has been a growing schism within urban and regional planning theory over the last ten years (Beauregard, 1989). Thus the technical and administrative techniques created in the past are seen as being based on a paradigm of narrow scientific rationalism, which is now widely regarded as being inappropriate. It now appears to be the turn of communicative planning (also argumentative and collaborative planning) (Fischer & Forester, 1993; Healey, 1996, 1997). This conception of planning accepts limits to power, to empirical knowledge, to the resolvability of moral dilemmas, but also seeks to enable the world-of-action to start out or move on toward something better, without having to specify precisely a goal. It is future seeking rather than future defining. Urban and regional planning theorists have long acknowledged that any planning framework embodies views and decisions about the nature of the socio-economic forces, the public and other interests and the planning objectives. But, since any planning framework exerts power over subsequent decisions, the preparation and use of regulatory frameworks in spatial planning are now commonly seen as more than a technical and bureaucratic exercise, and clearly involve political struggle. As a result, Throgmorton (1993) reflects that planners 'learn, and come to say, that planning and analysis are technical and disciplined by objective methods, but they also learn, and come to fear, that planning and analysis are political and subject to outrageous manipulations' (p.119). This realisation or acceptance permeates thinking in planning practice and the theorising about planning, as well as the area of concern in this paper, the nature and scope of evaluation.

One of the many reasons for this theoretical re-examination has been the apparent failure of the evaluation techniques and methods used in spatial planning. Shefer and Kaess (1990) claim that planners depend heavily on this interaction between objectives and evaluation methods in their analysis, planning and implementation processes. Reviews of such methods, (see for example Banister, 1994; Lichfield, 1996; Shefer and Kaess, 1990; and Voogd, 1997) reveal that many techniques have been developed over the last twenty-five years to serve the paradigm of scientific rationality. Thus in practice, cost-benefit analysis still remains in use, and is regarded as a viable method for clarifying the consequences of different choices despite the many fundamental shortcomings identified by Hill (1968) and Self (1975). The planning balance sheet (PBS) and goals achievement matrix (GAM) have also had an important impact, but as Voogd (1997) claims, not so much for the technical sophistication of their methods but more their power of conviction and transparency. These approaches have also remained in use, largely in their original form, since critiques have mainly focused on the technocratic method of their use. Multi-criteria analyses (MVA) developed in response to these criticisms in an attempt to take account of the conflicting interests and the many dimensions that planners have to arbitrate between.

However, more fundamental criticisms and views of urban and regional planning emerged in the 1980s. It was argued that the prevailing methods for evaluation were biased in favour of the status quo and ignored equity and participation issues. Planners became more aware that public decisions redistribute costs and benefits between different interests. Thus Lichfield (1996) has further developed the PBS model to construct a process of community impact evaluation (CIE). He claims that -

> ...there is no completely scientific objective means of striking a balance between environmental and other conditions.....Accordingly, the reasons for a decision should be open and accountable, and the value judgements that necessarily underlie it should be clearly identified, and there should be the widest possible opportunity for others who may be affected to contribute to the decision. (p.75)

But he acknowledges that CIE is rooted in the old rationality constructs of impact assessment and cost-benefit analysis. The concern with sustainable environments (Pearce, 1993) has also seen the development of further methodological advances in the measurement of environmental costs and benefits within Environmental Impact Assessments (EIA). Although Voogd (1997), for example, claims that there is not much evidence that EIAs play an important role in actual decision making in the Netherlands.

Finally, in many areas of public policy evaluation, scientific rationality is seen to be challenged by an economic rationality. As Healey (1994) claims –

> public policy in Britain has seen the increasing penetration of the vocabulary of economic evaluation into public policy formulation and implementation...a methodology of neo-liberal political ideology...an attempt to establish a dominant hegemony which crowds out the voices of other systems of meaning, while privileging big capital. (p.43)

Indeed, it is acknowledged that the use of performance targets, indicators and output objectives alongside new contracting arrangements for services which are inherent to this hegemony (Henkel, 1991), is an experience which extends well beyond our island, and has been referred to as a new public management (Pollitt, 1995). Although not an approach widely applicable to or used in urban and regional planning apart from setting targets for service delivery, it has permeated other areas of spatial policy in the UK, such as the evaluation of regeneration policies (Hambleton and Thomas, 1995; Oatley, 1998).

Thus in concluding their review of evaluation methods in urban and regional planning, Shefer and Kaess (1990) suggest evaluation practice has echoed the prevailing social climate, and despite the array of methods available, practitioners have remained conservative in their trends. Voogd (1997) goes further and believes that traditional concepts that underpinned evaluation practice, such as comprehensive rationality, have been replaced by new concepts such as consensus building, and highlights a growing need for a more open, but well-structured decision-making process. Interestingly, he propounds that evaluation methods should be adjusted to incorporate negotiated knowledge, believing this to be an interesting new avenue to explore by academic planners. Whilst this call has been taken up, Faludi & Altes (1994) suggest that the proponents of communicative planning give no or few clues on how evaluation might occur. Instead, they argue this new form of planning 'focuses on what planners do in their day-to-day work...with an emphasis on how they can promote democratic values' (Faludi & Altes, 1994 p.406). The evaluation role for the planner or planning in these concepts is thus expressed in terms of enabling activities, of rhetorical criticism and reflection. These represent very complex arguments concerning the nature of the communicative planning process, and as yet do not present clear normative views on practical courses of action.

In proposing new forms of communicative planning, Healey (1994, 1996) argues for the development of inter-discursive policy formulation largely based on the work of Habermas. This would address the need for empowerment, the demands of multi-cultural societies, and the cleavage and differentiation in society's interests by establishing new forms of communication.

> This involves not merely assessing who wins and loses with respect to particular issues, but whose terms dominate the discussion and who is included and excluded by this. The normative challenge is to invent forms of inter-discursive communication...which also have the power to resist or at least limit the discursive domination of powerful groups in plan-making and subsequent use. (Healey, 1994, p.45).

But as Healey (1994) herself reminds us - how far is it possible to imagine that a development plan can be anything other than either a project of the powerful or an ineffective dream of the idealistic? Further, as Karlsson (1996) argues:

> trying to solve questions about criteria through a power-free dialogue between interest groups is associated with Habermas...the argument against such a strategy for solving conflicts between different interest groups is that it appears to be too idealistic...Habermas's theory of dialogue contains within it the dream of an ideal form of life, without any ruling elite... (p.411)

Instead, Rebien (1996) sees relationships of power to be involved in all processes that have to do with knowledge creation, knowledge communication and knowledge use. Thus following Foucault's line of thinking, Rebien (1996) suggests that power should be looked at as something natural and productive. Consequently, from this perspective, although the power relationships are asymmetrical, this does not prevent interaction between stakeholders, it may be restricted but not powerless.

The creation of knowledge is inherent to the concept of a learning process, which has been incorporated in much of the evaluation debate. It is usually seen as some form of critical dialogue, which allows a more probing and accountable process of government actions to emerge (for example see Laughlin and Broadbent, 1996), a similar role as that seen for the rhetorical planner. Throgmorton (1993) argues that

...planning analysts should think of survey research and other tools as rhetorical tropes that reply to prior utterances (and give meaning and power to the larger narratives of which they are part), seek to persuade specific audiences, create open meetings subject to diverse interpretations, and help to constitute planning characters and communities... They should embrace persuasive discourse and political conflict and realise that survey results are, like all alleged 'facts' of planning analysis, inherently tropal and contestable... Surveys must be scientific and rhetorical, professional and political, because they, like all other planning and analytical tools, configure policy-oriented arguments. (p.133)

In a similar way, Karlson (1996) sees the critical dialogue in evaluation as the goal of understanding. In his view, the goal is to reach a greater insight and clarity of the foundations upon which one's own and others judgements are based, and by that to become more enlightened and active in participant in evaluation. The evaluator is seen as a provocative questioner, whose role is not to develop thought but to break up thought. Crucially, he also recognises that it is also important to acknowledge that the dialogue strategy is not an unproblematic one. The difficulty with dialogue as a strategy is that it demands that each interest group has enough resources to be able to participate. There is a risk of only those who are resource-powerful achieving participation.

Drawing conclusions from this debate is difficult, and obviously the theoretical underpinnings of evaluation need further development and exploration. A task we would argue as being particularly useful within the current dialogue in planning theory. Ultimately, we would also argue for the practice of evaluation to also be assimilated within these discourses. We support the claim by Radaelli and Dente (1996) that evaluation can be described as a 'process of transformation of ideas and knowledge' (p.59) and the transmission of knowledge into the policy process is seen to be a crucial issue. Within the current advancement of communicative planning, this raises the question of how to develop an evaluation that gives the stakeholders, especially those who lack power, a better chance to have their say (Healey, 1994; Karlsson, 1996). Whilst models based on scientific rationality, either in the form of welfare economics and systems analysis or the new public management, may be seen as excluding hegemonies, the search for an alternative means of dealing with the twin concepts of power and knowledge still remain deeply contested, but cannot be denied.

We therefore have to ask of any evaluation -

• who carries out the evaluations and who participates?
• at what level of governance are they carried out and how transparent are they?
• what is the normative content, the knowledge base and the knowledge-boundaries of the evaluations?

We now wish to consider how the debates and tensions between alternative paradigms of rationality can be used to probe at another new construct, the emergent activities of European-wide spatial planning.

Evaluation of Transport Networks and Structural Funds

Above we suggest that the role of evaluation is a live and contested issue at the heart of the EU spatial approach. The contested territory includes the selection of instruments/tools for evaluation, and the embedding of these tools in evaluation frameworks. As a way of illustrating these issues, we attempt to outline how evaluation is used in two areas of sectoral policy which will be key elements of future EU spatial planning: the trans-European transport network (T-TEN), and the Structural Funds.

A major concern in the development of a policy framework for T-TEN, has been the selection of appropriate evaluation techniques in the planning of projects, corridors, and the network as a whole. A line of cleavage in this struggle opened between the advocates of cost-benefit analysis (CBA), and the advocates of strategic environmental assessment (SEA). Richardson (1997) observes that whilst the case for economic rationality has been put consistently and strongly by many actors in the process, the call for alternative evaluation came from the margins. It was perhaps not surprising that economics should be the appropriate discipline to evaluate measures that were designed principally to bring economic benefits. However, the increasing concern over the environmental impacts of T-TEN precipitated a campaign for broader impacts and options to be considered in making decisions at network and corridor levels (Richardson, 1997). The construction of Strategic Environmental Assessment (SEA) and corridor analysis was weakened by a number of factors resulting from political processes. So, for T-TEN, a heated political struggle over the selection of evaluation instruments was central to the development of policy. The attempts to shape and control the evaluation process can be understood as struggles over the embedding of particular values, knowledge and power relations. The adoption of particular approaches and techniques in evaluation has created boundaries of inclusion and exclusion of knowledge, which potentially skews the outputs of evaluation. A range of evaluation tools is emerging with a mix of roles. Macro- and micro-economic evaluation is used to justify EU financial support for individual transport corridors, and to justify the EU's broad support for the T-TEN. Environmental evaluation tools (SEA and corridor analysis) will provide information on the overall impact of the networks, and different modal options at the corridor level. The outcome of a paradigmatic battle appears to have been the elevation of economic evaluation to hegemonic status. Economic criteria will be used to justify EU intervention in projects, whilst environmental knowledge will support decisions rather than carry any binding power. The value of international mobility will be more significant in influencing decision-making than the value of environmental impacts (Richardson, 1997).

Power is also found to be a crucial consideration in the process of evaluation. The hegemony of economics in evaluation would appear to favour the wide range of interests which are likely to benefit most from the construction of the trans-European networks. Concerns about the environmental impacts of the networks are marginalised. The particular methodologies further compound this exclusion of interests. CBA is a technical exercise which does not lend itself to the participation of multiple actors. Its values and workings are not transparent. SEA, alternatively, can be implemented readily within a participative planning framework. However, in the case of T-TEN, it does not appear that this possibility is being followed - rather SEA, like CBA, is being carried out as a desk exercise (Richardson, 1997). The processes by which these methodologies have been selected and incorporated into a policy framework are opaque, and do not benefit from public participation. The ground rules are being laid down for projects which will be implemented within member states, yet evaluation is not being used in a way which opens up the planning of projects, corridors and networks. It appears that innovations in evaluation in the T-TEN policy process will provide information about the broader environmental impacts of infrastructure development in major international corridors and networks as a whole, but will not create transparent processes of decision-making, or broaden the scope of the policy debate.

The nature of decision-making also lies at the heart of the Structural Funds, which by 1999 will have invested a total of 200 billion Ecu in the assisted regions of Europe. It was widely acknowledged that the monitoring and control of community regional expenditure was inadequate prior to the 1988 reform of the Structural Funds (McEldowney, 1992), and at the time of the Maastricht Treaty the European Council had to request that greater emphasis should be given to monitoring and ex-ante and ex-post evaluation (Bachtler and Michie, 1995). Apart from the principle of better accountability per se, the evaluation issue was given added urgency in 1993 by the much greater volume of Community resources being allocated to Structural operations over the 1994-1999 period, as the profile of economic and social cohesion in the EU was increased. A key issue for the evaluation of the Structural Funds has been the joint funding of Member State projects, and since 1988, Member State regional programmes. There have been different expectations and traditions of evaluation between Member States, and the lack of detailed guidance from the Commission meant consensus was lacking over the objectives of evaluation and the methodologies to be employed. The Commission was to give a much clearer and co-ordinated view on evaluation after Maastricht. A clear distinction was drawn between three stages of Structural Fund assesment- appraisal, monitoring and evaluation. Bachtler and Michie (1995) indicate that although some sophisticated model-based approaches have been used, the majority of EU evaluation studies have tended to favour a combination of macro-level research involving before and after numerical analyses of a limited range of socio-economic indicators, and micro-level, largely non-experimental surveys involving the collection and basic analysis of a mix of qualitative and quantitative information through interviews, expert views and case studies. In spite of greater consensus in the approaches, the evaluation of Structural Fund operations still presents a host of frequently cited methodological problems (Bachtler and Michie, 1995; McEldowney, 1992, 1997), but it has been the exercise of power, rather than the nature of knowledge which has been the most contested area of Structural Fund assessments.

It was the duty of competent authorities in the Member States to ensure that appraisal and evaluation were carried out in the most effective manner. This recognised the principle of subsidiarity in implementation, but also allowed national evaluation techniques and approaches to determine the nature of evaluation. The European Commission, in doing this, was seeking to emphasise and utilise the important notion of partnership, a feature central to the operations and objectives of the Structural Funds. Towards this end, the Commission established a specialist evaluation unit and invested in research on methodologies, the MEANS programme, to develop a European evaluation culture. The general aim was to provide the Commission with a coherent body of evaluation guidelines and methods and particularly adapted to partnership management. According to Monnier (1997), vertical partnerships constitute a favourable context for the development of evaluation, and provide the means to carry out quality evaluations since situations of partnership favour pluralistic approaches. The outstanding feature of the pluralistic approach is that it takes into account the different points of view of people involved in the policy process. Partnerships are also seen to provide an opportunity for stakeholders' involvement and to enhance the quality of the process of evaluation. Monnier (1997) acknowledges that partnerships introduce a new series of constraints. Thus, one of the main challenges facing evaluations of activities implemented in partnership is to produce recommendations which, if not consensual, are at least seen by all partners as being credible (a multijudge-multicriteria technique).

In practice, the 1993 Regulations have only seen the ex-ante appraisals undertaken to date, but more generally the preparation of regional development programmes has frequently been seen as a highly confrontational exercise between the Commission and the Member States, or as in the UK, between levels of government within the Member State (Lloyd, 1996). As Eskelinen et al (1997) claim, although the subsidiarity principle is emphasised in regional development programming, the supranational institutional practices often set the tone of Structural Fund operations. This concerns the diagnosis of the problems and the strategies for their alleviation, which necessarily implies the issue of compatibility of European interests with national and regional interests. Often, these interests have been served by evaluations which support the overriding hegemony of integration, based on economic cohesion more than social inclusion, and increasingly driven by fiscal restraint in the search for Monetary Union, as exemplified by SEM 2000 (Liikanen, 1997) and more broadly the new public sector management adopted in many Member States (Pollitt, 1995). However, Eskelinen et al (1997) suggest that evaluations should not be considered merely as a task that the Member States have to 'carry out for Brussels' (p.172), nor be primarily regarded as a way of securing the legitimisation of the programmes. At best, evaluations might contribute to new forms of interaction between research and governance, and thus influence the planning and implementation of regional policy measures. But as Bachtler and Michie (1995) conclude, in a highly politicised context, with fifteen Member State administrations and a multiplicity of regional partners, the Commission's efforts to establish a commonly appreciated framework continues to be an uphill struggle.

The importance of these skirmishes is that they represent a series of struggles over the nature of knowledge as the basis for policy making, and over the dynamics of power in the evaluation process. This can be understood as constructing a new basis of rationality or legitimacy for policy in Europe. Elsewhere, and more broadly, Giannakourou has argued that the construction of rationality in public policy making at the EU level is being shaped by broader socio-political discourse. In particular, in the development of EU spatial planning, principles of spatial equity have been subordinated to the over-riding needs of market integration (Giannakourou, 1996). However, the argument continues, an alternative locus of equity has emerged, which may be more appropriate to the European market: the emergence of a new inter-institutional, pluralist model of policy making. Procedural equity now replaces spatial equity, with a new emphasis on horizontality rather than hierarchy. This can be seen in the partnership principle which applies to the structural funds and the shared responsibility principle for environmental policy. The logic and challenge here is that the emergence of EU spatial planning offers opportunities to explore new types of rationality, in a shift away from the comprehensive evaluation that has characterised national planning systems. Whilst we would accept that this shift is conceptually in line with a broader search for forms of governance beyond the state, and would welcome the challenge to adopt a positive role in constructing rationalities, we are less convinced about the extent of this shift in practice, and believe that the nature of this claimed equity warrants close examination.

Concluding Comments

The experience of evaluation in spatial policy is clearly critical to the development of the European spatial approach. As transnational planning takes shape, a major issue for planners (theorists as well as practitioners) will be to establish methods of planning which will reach across contrasting planning systems, and find some balance/synergy between the aims of each member state as well as the wider community. In this construction site, we will need to be critical aware of these dynamics, and their political sensitivity.

In the case of T-TEN, the evaluations in many instances are limited in their use as planning tools. They are usually indicative rather than binding, but constrained in their scope, and reductionist in nature. As constructed in the T-TEN policy process, even SEA, which offers great potential as a participative planning tool, fails in this respect. So in the development of EU spatial planning, there are several difficult challenges to be overcome if the planning framework is to do more than simply legitimise and enable the policies and visions of the single market. A central question, then, will be how are tools to be developed which really function as planning tools - in their methodology, normative content, transparency and deployment, rather than simply serve to legitimate and expedite hegemonic visions, strategies and policies.

To accommodate the shift to pluralism, to whatever extent this occurs, evaluation methodologies will need to adapt to new types of policy processes, and be grounded in alternative rationalities. However, the rationality of the policy framework for T-TEN appears to be very much the old scientific rationality, whilst in the Structural Funds, a trend towards new public management can be seen. There seems scope here for experimentation with approaches to evaluation. A major challenge remains, though, which marks the distinction from evaluation in other areas. This is how spatial equity can be evaluated and then integrated effectively into planning policy through evaluation.

This leads us to a fundamental question. If pluralism is to become the model for evaluation, replacing traditional hierarchical approaches, then it follows that the values which become the basis for evaluation are arrived at through some pluralist process. What is absolutely critical in this process, then, is the extent of inclusion or exclusion of particular values. In a market-oriented paradigm, we may expect that the central values are likely to follow the needs of the market. Such values would be expressed through the multiple actors representing different interests. The inclusion of particular actors would thus shape the normative content of evaluation, determining the boundaries of the knowledge base, the scope, and potentially the outcomes of evaluation. In the T-TEN policy process, the nature of partnership has maintained a narrow scope for evaluation and policy development, which suggests caution in the partnership approaches being explored in the Structural Funds. Of particular relevance to the EU spatial planning is the subsidiarity principle, which recognises the need for transparency of the decision-making process as a critical factor in strengthening the democratic nature of the EU institutions, and supporting the public's confidence in the administration. We conclude that reflections on the nature and use of evaluation are important to European spatial planning, and are urgently needed. There is a need for empirical studies and evidence from evaluation practice, at the national and European levels, which explore evaluation in planning in the context of emerging alternative theoretical paradigms.

Note

1.   Earlier working versions of this paper were presented to the XI AESOP
        Congress, University of Nijmegen, The Netherlands, in May 1997, and at the
        Regional Studies Association EURRN Conference, at the European Universitat,
        Frankfurt Oder, Germany, in September 1997

References

Bachtler, J. & Michie, R. (1995), A new era in EU regional policy evaluation ? The
        appraisal of the structural funds. *Regional Studies,* 29 (8), 745-752.
Bachtler, J. & Turok, I. (eds.) (1997), *The coherence of EU regional policy : contrasting
        perspectives on the structural funds.* London: Jessica Kingsley Publishers.
Banister, D. (1994), *Transport planning.* London: E&FN Spon.
Beauregard, R. (1989), Between modernity and postmodernity : the ambiguous position of
        US planning. *Environment and Planning D : Society and Space,* 7, 381-395.
Carley, M. (1980), *Rational techniques in policy analysis.* London: PSI.
CEC (1994), *Europe 2000+: co-operation for European territorial development.*
        Luxembourg : Office for Official Publications of the European Communities.
CEC (1997a), *The EU Compendium of Spatial Planning Systems and Policies,*
        Luxembourg: Office for Official Publications of the European Communities.
CEC (1997b), *European spatial development perspective : first official draft.* Presented at
        the Informal Meeting of Ministers Responsible for Spatial Planning of the
        Member States of the European Union, Noordwijk, June 1997. Luxembourg:
        Office for Official Publications of the European Communities.
CEC (1998), *European spatial development perspective : complete draft.* Presented at the
        Informal Meeting of Ministers Responsible for Spatial Planning of the Member
        States of the European Union, Glasgow, June 1998. Luxembourg: Office for
        Official Publications of the European Communities.
Chen, H-K. (1990), *Theory-driven evaluations.* London: Sage.
Committee for Spatial Development (1996), *Spatial planning,* Discussion document on
        spatial planning presented by the Italian Presidency for the Ministerial Meeting
        on Regional Policy and Spatial Planning, Venice, May 1996.
Davies, H. W. E. (1994), Towards a European planning system?, *Planning Practice and
        Research,* 9 (1) 63-69.
Department of the Environment (1992), *Evaluating the effectiveness of land use planning.*
        London: HMSO.
Eskelinen, H., Kokkonen, M. & Virkkala, S. (1997), Appraisal of the Finnish Objective 2
        Programme : reflections on the EU approach to regional policy. *Regional Studies,*
        31(2), 167-172.
European Commission (1997), *European spatial development perspective : first official
        draft.* Luxembourg: Office for Official Publications of the European
        Communities.
Evans, B. (1997), From town planning to environmental planning. In Blowers, A. &
        B.Evans (eds.) *Town planning into the 21st century.* London: Routledge.
Faludi, A. (1997), European Spatial Development Policy in 'Maastricht II' ? *European
        Planning Studies,* 5(4), 535-544.

Faludi, A. & Altes, W. (1994), Evaluating communicative planning : a revised design for performance research. *European Planning Studies* , 2(4), 403-418.

Fischer, F. & Forester, J. (eds.) (1993), *The argumentative turn in policy analysis and planning*. London: UCL Press.

Giannakourou, G. (1996), Towards a European spatial planning policy: theoretical dilemmas and institutional implications, *European Planning Studies,* 4 (5) 595-613.

Greed, C. (ed.) (1996), *Implementing town planning*. Harlow: Longman.

Guba, E. & Lincoln, Y. (1989), *Fourth generation evaluation*. London: Sage.

Hague, C. (1996), Viewpoint: Spatial planning in Europe: the issues for planning in Britain, *Town Planning Review*, 67 (4) iii-vi

Hambleton, R. & Thomas, H. (eds.) (1995), Urban Policy Evaluation : challenge and change. London: Paul Chapman Publishing.

Healey, P. (1994), Development plans : new approaches to making frameworks for land use regulation. *European Planning Studies* 2(1), 39-58.

Healey, P. (1996), The communicative turn in planning theory and its implication for spatial strategy-making. *Environment & Planning B,* 23, 217-234.

Healey, P. (1997), *Collaborative Planning*. London: Macmillan Press.

Henkel, M. (1991), *Government, Evaluation and Change*. London: Jessica Kingsley Publishers.

Hill, M. (1968), A goals-achievement matrix for evaluating alternative plans. *Journal of the American Institute of Planners* 34(1), 19-28.

Hull, A. (1996), Strategic plan-making in Europe: institutional innovation, *Planning Practice and Research*, 11 (3) 253-264.

Karlsson, O. (1996), A critical dialogue in evaluation : how can the interaction between evaluation and politics be tackled ? *Evaluation,* 2(4), 405-416.

Laughlin, R. & Broadbent, J. (1996), Redesigning Fourth generation evaluation : an evaluation model for the public-sector reforms in the UK. *Evaluation,* 2(4), 431-452.

Lichfield, N. (1996), *Community impact evaluation*. London: UCL Press.

Liikanen, E. (1997), Evaluation in the European Commission : an interview with Commissioner Liikanen. *Evaluation,* 3(2), 237-244.

Lloyd, P. (1996), Contested governance : European exposure in the English Regions. In Alden, J. & Boland, P. (eds.) *Regional Development Strategies - a European Perspective*. London: Jessica Kingsley Publishers.

McEldowney, J. (1992), Evaluation and European regional policy. *Regional Studies,* 25(3), 261-265.

McEldowney, J. (1997), Policy evaluation and the concepts of deadweight and additionality: a commentary. *Evaluation* 3(2), 175-188.

Monnier, E (1997), 'Vertical' partnerships: the opportunities and constraints which they pose for high quality evaluations, *Evaluation,* 3(1), 110-118.

Morphet, J (1996), Lessons from Europe:1, *Proceedings of the Town and Country Planning Summer School*.

Morphet, J. (1997), Enter the ESDP - plan sans fanfare. *Town & Country Planning* 66(10), 265-267.

Oatley, N. (ed.) (1998), *Cities, economic competition and urban policy*. London: Paul Chapman Publishing.

Pawson, R. & Tilley, N. (1997), *Realistic evaluation*. London: Sage.

Pearce, D. (1993) *Economic values and the natural world.* London: Earthscan Publications.

Pollitt, C. (1995), Justification by works or by faith? Evaluating the new public management. *Evaluation 1(2),* 133-154.

Radaelli, C. & Dente B. (1996), Evaluation strategies and analysis of the policy process. *Evaluation* 2(1), 51-66.

Rebien, C. (1996), Participatory evaluation of development assistance : dealing with power and facilitative learning. *Evaluation,* 2(2), pp.151-173.

Richardson, T. (1997), The Trans-European transport network : environmental policy integration in the European Union. *European Urban and Regional Studies,* 4(4), 333-346.

Self, P. (1975), *Econocrats and the policy process ; the politics and philosophy of cost-benefit analysis.* London: Macmillan.

Shaw, D., Nadin, V., and Westlake, T. (1995), The Compendium of European Spatial Systems, *European Planning Studies,* 3 (3), 390-395.

Shefer, D. & Kaess, L. (1990), Evaluation methods in urban and regional planning. *Town Planning Review* 6(1), 75-88.

Throgmorton, J. (1993), Survey research as rhetorical trope : electric power planning arguments in Chicago. In Fischer F. & Forester J. (eds.) *The argumentative turn in policy analysis and planning.* London: UCL Press.

Voogd, H. (1997), The changing role of evaluation methods in a changing planning environment : some Dutch experiences. *European Planning Studies* 5(2), 257-266.

Williams, R. H. (1996), *European Union spatial policy and planning,* London: Paul Chapman Publishing.

8 The Development Planning Process in Six English Regions

MARK BAKER

Introduction to the English Development Planning Process

Land use planning as a governmental activity can be undertaken at a variety of spatial scales. Within the UK, it is possible to identify at least four such levels from that of the nation-state downwards and, in addition, recent years have also seen the emergence of an international dimension involving the entire European Union and sub-regions within it (Figure 8.1). The balance of power and responsibilities between these different spatial scales does, however, vary from country to country and over time. Within the UK, spatial planning at the regional level has often appeared to be weaker than in several other European states, although it is still possible to identify periods when regional policy development did became more prominent (Wannop & Cherry, 1994). Most notable amongst these was that of the early-mid 1940s which witnessed the a so called 'classic era' of regional planning, typified by Abercrombie's famous plans for Greater London (1944) and the Clyde Valley (1946). These exercises in regional thinking strongly influenced the subsequent development of national policy and its local implementation (today's metropolitan green belt, radial motorway and new towns, for example, were central features of Abercrombie's visions for Greater London) but the passing of the 1947 Town and Country Planning Act, which heralded the beginnings of a national planning system, effectively bypassed this regional scale by introducing statutory development plans at the local authority level without an explicit provision for any intermediate tier of regional policy formulation. Central government, however, ensured continued implementation its own national policies by requiring such development plans to be formally approved by the Ministry.

Spatial Level	Geographical Area	Institutional Framework	Land-use Policy Related Outputs
International	European Union and European Sub-Regions	European Parliament & Commission	EU Directives; European Spatial Development Perspective (ESDP)
National	UK; England	Parliament; DETR	National Legislation; National Planning Policy (PPGs)
Regional	English Regions	Government Offices for the Regions (GORs); Regional Development Agencies (RDAs); Regional Chambers	Regional Planning Guidance (RPGs); RDA Economic Strategies
Sub-Regional	Counties; Metropolitan Areas	County Councils; Sub-regional Associations of Unitary Councils	Structure Plans; Strategic Planning Guidance (SPG)
Local	Districts	District and Unitary Councils	Local Plans; UDPs

Figure 8.1 Examples of different spatial levels of land-use planning

Although the Labour administrations of the 1960s and 1970s did re-introduce a regional dimension to land use planning, and particularly economic development, these were abolished by the incoming Conservative regime in 1979. What did, however, endure from the late 1960s was a re-vamped development plan system which consisted of a mandatory strategic element (structure plan) at the sub-regional (county) level and powers to prepare more detailed, discretionary local plans at the local level. The move to a two-tier structure of local government (counties and districts / metropolitan boroughs) in the early 1970s effectively separated strategic and local planning responsibilities between county and district authorities. Although the abolition of the greater London Council (GLC) and the English metropolitan counties in the mid-1980s re-introduced a single planning policy document (the unitary development plan) into the major English cities at the expense of conurbation-wide strategic thinking (Roberts et al, 1999). In 1992, the preparation of district-wide local plans became a mandatory requirement, although a significant number of local authorities have still not met this responsibility. Most recently, further changes to local government structures in the 1990s have also seen the introduction of unitary tiers of local government in many shire areas of Britain but, with the main exception of Wales, a two-tier (structure and local) development plan system has been retained.

Despite these structural changes outlined above, the essential features of the English planning system are, however, in many ways little changed from those introduced over 50 years ago with the passing of the 1947 Act. The system is essentially an administrative and regulatory one with two key components: planning permission is required for the development of land, and local authorities are responsible for preparing development plans which set out policies and proposals on the future use of land in their areas and provide some guidance for the determination of planning applications. As far as this statutory system is concerned, the spatial focus of powers and responsibilities are thus concentrated at the national and local levels. Overall responsibility for the operation of the planning system in England rests with the Secretary of State for the Environment, Transport and the Regions, but detailed development control decisions and the formulation of development plan policies are generally local government responsibilities. Nevertheless, central government retains reserve powers of intervention in respect of both local development plan content and individual decisions.

The Secretary of State also issues *national planning policy* which has legal status as a 'material consideration' in planning decisions, as well as more directly setting a context for development plan preparation. Today, the principal source of such national planning policy is through the planning policy guidance (PPG) series. The first PPGs were introduced in 1988 and, in line with the ideology of the Thatcherite administration of the time, generally took a market-oriented stance. Different approaches, embracing environmental and sustainability perspectives, have increasingly been incorporated into new and revised PPGs throughout the 1990s. But, at the same time, the amount and scope of such guidance has also expanded rapidly with the result that, by the mid-1990s, there were over twenty PPGs covering a wide variety of land-use topics. During the same period, new forms of *regional planning policy* began to re-emerge via the regional planning guidance (RPG) series. Nevertheless, this was a far cry from the earlier land use planning exercises of the 1940s and 1970s. Despite their regional focus, RPGs are also issued by the Secretary of State although the early stages of the preparation process do involve the provision of advice on content by regionally-based associations of local authorities. Since its inception, this RPG series has been roundly criticised, not only for the degree of central government control over its preparation, but also for its extremely limited and anodyne content which has tended to merely repeat existing national policy without the addition of genuinely regional perspectives (Roberts, 1996; Baker, 1998). Although changes introduced by the current Labour central government (DETR, 1999) appear to address some of these criticisms, it remains much less certain whether a regional level of governance will emerge over the next few years as a major challenge to existing centres of power at the national and local levels.

Local Development Planning in a Plan-led System

It is therefore against this backdrop of extensive national planning policy, and centrally influenced regional policy, that local authorities are responsible for preparing their *local development plans*. Although the British planning system is often termed as a 'discretionary' because the content of statutory development plans are not legally binding on subsequent development decisions, the relationship between the policies within adopted development plans and subsequent planning decisions was re-emphasised in the early 1990s with the passing of the Planning and Compensation Act 1991 which introduced the now-famous additional 'section 54a' into the earlier Town and Country Planning Act, 1990 as well as making the preparation of district-wide local plans a mandatory responsibility of all shire district councils. Although initial government targets of complete plan coverage by the end of 1996 proved over-optimistic, the latter provision should ensure that the whole of England will eventually be covered by detailed planning policies.

At the time, these legislative changes were heralded by central government as establishing a clear commitment to a 'plan-led' system of development control. There has subsequently been a great deal of debate in academic, legal and professional practice arenas over the true significance of these legislative changes (particularly section 54a) to the day-to-day operation of the planning system. A number of academic papers and research reports have been published which have mainly focused on the role of development plans in the determination of planning applications and appeals and have therefore mainly followed a legal perspective (eg. Gatenby and Williams, 1996; Purdue, 1994; MacGregor and Ross, 1995). Attitudes vary from those who regard the 1991 Act as the greatest change to planning law since 1947, to others who argue that nothing much has altered at all. However, recently revised national planning policy (DoE, 1997) still continues to emphasise that development plans are at the heart of the planning framework and are considered by government to be the most effective way of reconciling demand for development and the protection of the environment.

In addition to the widespread interest in the impact of this new legislation on decision-making in development control, the moves towards a plan-led system can also be considered within a more general context of public administration and central/local government relationships as, on the face of it, the new system appeared to represent a shift in decision-making power from the centre towards local planning authorities responsible for preparing development plans. Certainly, the changes were initially seen in this light by some local authorities who had felt marginalised during the mid to late 1980s as the market-led, anti-interventionist policies of the Thatcher era were translated into a planning regime which tended to bypass local government and become characterised by ad hoc decision making via planning appeals. With the introduction of the 'plan-led' system, local control over decision-making was potentially enhanced by a more rigorous implementation of local plan policies. At the strategic level, the shift of power towards the locality appeared to be further reinforced by another provision of the 1991 Act which removed the previous requirement for structure plans to be approved by the Secretary of State for the Environment before their adoption.

A few commentators did, however, take a different stance, arguing that although the legislative changes may well have strengthened the role of locally-formulated development plans to some degree, the parallel moves towards increased national and, to a lesser extent, regional planning policy have effectively constrained the content of such plans since, although termed 'guidance', both the PPG and RPG series are essentially policy and procedural documents which local planning authorities are required to take into account when preparing their development plans. Thus, Tewdwr-Jones (1994a; 1994b) suggested that the English planning system of the early to mid-1990s was characterised by the introduction of a planning policy hierarchy, operating 'from the top downwards' which is:

> ... too constrained by central government in both policy formulation and policy implementation .... decision-makers are starting to realise that the newly-found status of the development plan as a means of determining applications is off set by numerous checks and balances imposed by the Department of the Environment, one of which is referring to the contents of national policy documents (Tewdwr-Jones, 1994b, 592).

With hindsight, it does appear that many local authorities might have been, at least initially, blinded by the glare of the re-discovered importance of their development plans and therefore remained somewhat unaware of the degree of central control still inherent within the planning process. The key arena in which these conflicting tendencies towards decentralisation of decision-making (in the form of the increased status of development plans) and greater centralisation (through increased national and regional policy constraints), thus come face to face with each other is during the development plan making process. The outcome depends on the extent to which central government is able and willing to *intervene* in the local plan production process to ensure that national and regional policies are upheld. To this end, an examination of the potential means by which the centre can ensure compliance with national and regional policy, reveals a number of formal and informal channels throughout different stages of the plan-making cycle (Figure 8.2). Intervention may occur at any time during this process, although the form that this takes will vary: methods of intervention or, in a less formal sense, of influencing policy formulation include discussions between civil servants and local planning officers in the early stages of plan preparation; informal comments on published draft plans; formal objections to deposited plans; and, in the last resort, formal directions to modify plans or even call-in.

Central government was not, itself, slow to realise the importance of such intervention in the plan-making process in the immediate aftermath of the 1991 Act. Ministerial statements of the time emphasised the intention to adopt a more rigorous approach to the scrutiny of emerging plans and a willingness to intervene where necessary. In his opening address to the Town and Country Planning Summer School in September 1992, Sir George Young (Minister for Housing and Planning) stated that the implications of section 54a were not confined to local authorities, but had implications for the work of the DoE as well:

> We shall not hesitate to intervene if plans significantly conflict with national and regional policies and we are not satisfied that such conflicts are justified by local circumstances. Although we would naturally hope to resolve any difficulties at pre-deposit stage, the Department is quite prepared to object at deposit stage, if need be. And, of course, in the last resort there are reserve powers of direction and call-in. (Sir George Young, 1992, 4).

Stage of Preparation	Characteristics	Intervention
Pre-deposit	Earliest stages of plan production. Format left to discretion of local planning authority, but usually includes background surveys and research and the publication of some material for consultation purposes - often in the form of a draft plan (although no longer legally required).	Informal liaison with LPA and / or written comments on any published material.
Deposit	Plan formally placed on deposit for a statutory six week period during which representations and objections are invited from interested parties.	Formal objections by the Secretary of State.
Public Local Inquiry (PLI) or Examination in Public (EIP)	Opportunity for objections to be made in an open forum, chaired by an independent inspector. With structure plans, this is replaced by an EIP where the discussions cover selected topics and participation is by invite only.	Civil servants give evidence on behalf of the Secretary of State at PLI or EIP.
modifications	Following publication of the Inspector's (or EIP Panel) report, the local planning authority decides whether to modify any of the policies or proposals set out in the earlier deposit plan. If so, the changes must be placed back on deposit to allow an opportunity for objection. There can be more than one round of modifications if the LPA then propose to make further changes.	Formal objections by the Secretary of State.
adoption	Finally the plan is adopted, although there is provision for a six month period in which a legal challenge can be made. This happens rarely and can only cover issues such as procedural irregularities rather than further objections to policy content.1	Possible Directions to Modify, or call-in of plan*.

Figure 8.2 The plan-preparation process and opportunities for central intervention

* *The Secretary of State can make such an intervention at any stage in the preparation process, but it is usual practice to wait until close to the end of the process when all other avenues have been explored.*

The rationale behind such intervention was set out in the 1992 version of PPG1. This explained (DoE, 1992, para. 25) that the plan-led system, in effect, introduced a 'presumption in favour of development proposals which are in accordance with the development plan' and that the Secretaries of State have been examining development plans carefully to identify conflicts with national or regional policy guidance and would make formal interventions if necessary. Paragraph 29 of PPG1 went on to state that:

> ... if no such intervention is made, local authorities may take it that the Secretaries of State are content with the plan at the time of adoption and will attach commensurate weight to it in decisions they make on appeals or called-in applications.

The significance of this latter statement was to place the onus on the Department of the Environment to intervene in the plan preparation process in order to ensure compliance with national and regional policy. Failure to do so would mean that local policies which conflicted with national or regional policy but, nevertheless, became incorporated into adopted development plans would carry more weight in decision-making than national or regional policy (provided that the conflict has not arisen because of the publication of new guidance following the adoption of the development plan).

Central Intervention in Plan-making: A Survey of Six English Regions

Despite the copious amounts of academic literature on central-local relations in a general sense (eg, Loughlin, 1986; Rhodes, 1988; Stoker, 1991; John, 1994) and within the overall operation of the planning system and related areas of urban and environmental policy (eg. Brindley et al, 1989; Thornley, 1993), there has been little detailed academic research which specifically focuses on this crucial issue of central government influence on the development plan preparation process. This is despite evidence from the world of practice that some local authorities are increasingly questioning the extent of government involvement in their policy formulation (eg. Rosen, 1993; Long, 1995; Kitchen, 1997).

One recent exception to this paucity of detailed research (Ho, 1997), based on two case studies of Bolton and Islington, does explore the balance of power in central-local relationships in the plan-making process and comments on the possible undermining of both development plans and local democratic decision-making that central intervention can entail. Since both of these case study areas involved the relatively rare situation whereby the Secretary of State intervened with formal directions requiring modifications to the plans, these two chosen cases may be somewhat atypical of the more general situation elsewhere. Hull and Vigar (1998) have also considered central-local relationships in plan-making as part of a wider study of development plans and the regulatory form of the planning system. These case studies, predominantly focussing on strategic (structure planning) planning processes in Kent, Warwickshire and Lancashire, also encountered evidence of significant central intervention and feelings amongst local planning authorities that this had reduced the possibilities for local diversity in plan approach and expression. Interestingly, however, this view was not always mirrored by some developers and other private sector interests who felt that central government guidance was sufficiently vague to allow local authorities to follow their own agendas and that, in any case, the self-adoption process inherent in plan-making (such as Inspectors' recommendations not being binding) meant that they could ignore national policy anyway (Hull and Vigar, 1998).

The results presented below from another recent research project (undertaken by the author) also addresses these issues. However, in contrast to the mainly qualitative research already mentioned, this research was based on an more empirically-driven survey of all local planning authorities within six different regions within England[1]. The six regions covered in this research (Figure 8.3) can be split into two broad groupings of 'northern' and 'southern'. They thus incorporate a good spread of different types of local authorities and development plans and were intended to capture extremes in terms of economic conditions, demographic change, development pressures etc. (i.e. any 'north-south' divide). The overall aim of the research was to investigate the form and nature of central government's involvement in the plan-making process, including whether there was any evidence of regional differences in terms of the relationships between different Government Offices and local planning authorities, or in respect of the issues which frequently were subject to intervention. The outputs should, therefore, usefully complement the other studies mentioned above as well as introduce a firmer empirical basis to any subsequent research carried out by the author or others.

Comparing the mean number of objections recorded by the survey for each year between 1989 and 1995 (Figure 8.4) proves beyond reasonable doubt that civil servants did indeed take a more rigorous approach to scrutiny of plans after the introduction of the plan-led system

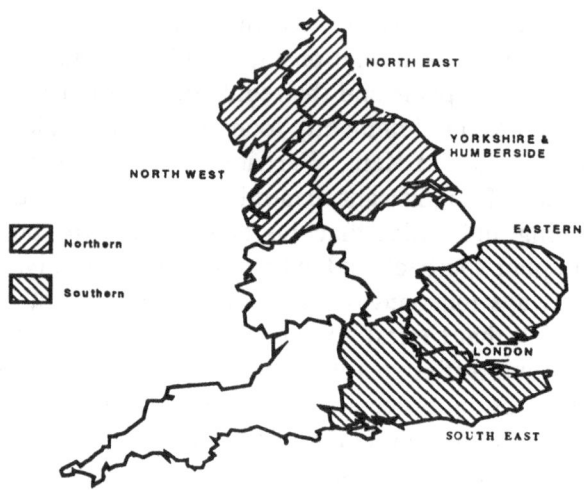

Figure 8.3 Government intervention in plan preparation: The six case study regions

Mean no. of objections

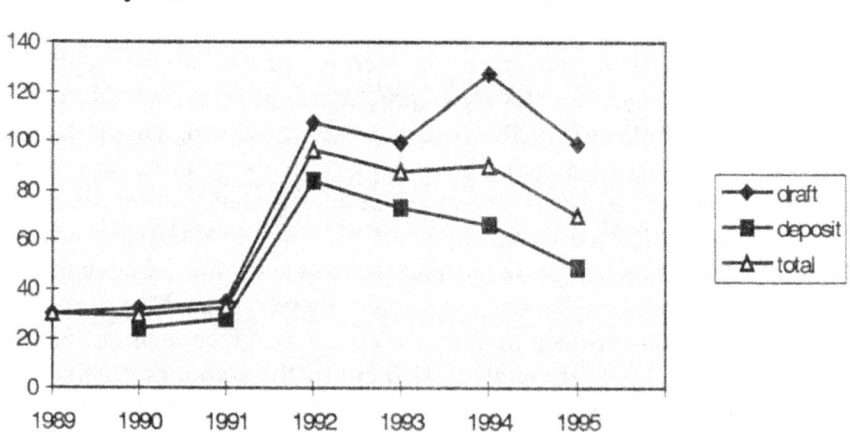

Figure 8.4 Central government objections per plan (mean) by year and plan stage

in 1992. The mean annual number of objections per plan increased threefold to not far short of one hundred per plan between 1991-92 before slowly declining again by 1995 to around seventy. The research thus confirms that the introduction of the plan-led system in 1992 was accompanied by considerably increased levels of central intervention in the preparation of development plans. Breaking the aggregated objections down into separate figures for plans at draft and deposit stages shows that both increased in a similar manner in 1992, but that comments at draft stage subsequently remained high, whereas the number of formal objections dropped. This is not, however, very surprising as increases in the number of comments at draft stage allows those local authorities that are willing to alter their plans to do so prior to deposit, thus reducing the need for formal objections later in the process.

This overall pattern of a rapid, post-1992 increase in the scale of intervention, followed by something of a tailing off, is generally mirrored in four of the six regions; the only exceptions being the Eastern region where the number of objections peaked somewhat later in 1994, and in London where a very different pattern can be explained by the number of plans which had already been scrutinised before 1992. Despite the increase in comments and objections, the usage of the most powerful reserve weapons of 'directions to modify' and call-in remained relatively rare – only ten cases of formal 'directions to modify' were recorded (representing under 15% of all plans which had reached adoption) and there were no 'call-in' cases. It is, however, quite possible that the mere existence of these powers ensures that local planning authorities do take significant account of earlier central objections, thus making the actual formal use of directions to modify and call-in unnecessary.

As well as making its views known at those formal stages which are set aside for open participation in the plan-making process, opportunities are also often available for central government policy-makers to influence emerging plan content via informal consultations and discussions outside the statutory consultation periods. The survey responses showed that the use of such informal channels was widespread in all regions, although particularly marked in the North East. Indeed, the results of the survey of both formal and informal intervention in the plan-making process suggested a deliberate strategy by policy-makers in the Government Office for the North East (GONE) to intervene both formally and informally in the early stages of plan production in an attempt to resolve problems before the need for stronger types of intervention at deposit stage and beyond. On the basis of the survey results, which showed the extent of formal intervention at later stages dropping off in the North

East as compared with most other regions, this approach would seem to have worked. Such informal mechanisms can, therefore, reduce the need for formal interventions at later stages, but also raise potentially serious questions about the subtle nature of central-local relations and the extent of decision-making that takes place behind closed doors.

Interestingly, the vast majority of recorded central government objections were seen to be non-topic specific, but instead relate to the operation of the planning system in a more general sense. Although such operational or procedural issues sometimes cover serious conflicts with national policy, the large number of seemingly minor 'wording' objections which cropped up in the survey provide ammunition to those (eg. Kitchen, 1997) who complain about the over-detailed nature of intervention in the plan-making process - indeed one respondent to the survey suggested a better title would be central government 'interference'! As well as introducing potentially unnecessary delays in the plan preparation process, another clear danger of such an central obsession with detail is that the resultant plans adopt a centrally-defined style and content, thus reducing the opportunity to reflect local or regional differences and incorporate a more locally democratic, people-oriented approach.

Where the objections were topic specific, conflicts with national policies on housing (including issues concerning housing allocations and affordable housing provision) were most numerous, followed by conflicts with green belt policy and that for the countryside and the rural economy. There were some north-south geographical differences evident, with housing issues identified as a particularly strong source of central-local conflict in the southern regions, whereas green belt and minerals issues are more common categories for objection in the north. At the regional scale, a few regions stood out as having particularly high numbers of objection in respect of certain policy areas. These included countryside objections in the Eastern Region and Yorkshire and Humberside (both of which contain extensive rural areas), nature conservation and coastal issues in the North East (presumably reflecting the international significance of the Northumberland coastline in particular), and planning and noise issues in the densely populated, urban metropolis of London.

Reflections on Central-local Relations within the Plan-making Process

Clearly there are arguments that central government has a legitimate role in upholding national planning policy (Quinn, 1996). In a cascading family of plans system as operates in England, there is a need for local policy to broadly reflect (unless particular local circumstances justify otherwise) overarching frameworks set at the strategic and national level. Nevertheless, local planning authorities are not merely implementation agents of central government. They also have their own legitimate roles in preparing and implementing local policies to address local problems, needs and opportunities. The question is therefore where to draw the line between granting local authorities complete freedom in policy adoption, and the imposition of what can amount to a uniform set of policies for all areas. One answer might be that the role of the centre should extend no further than matters of principle which could either undermine policies of genuine national significance, or adversely affect the overall operation of the planning system as a whole. However, the extent of current national guidance through the PPG series, coupled with the apparent willingness on the part of civil servants to become embroiled in minute details of policy formulation as revealed by the survey, appear to go much further than such necessary checks and balances.

Conversely, it could be argued that the relative paucity of topic-based conflicts revealed in the survey shows that local authorities have little desire to depart from national policy and there are therefore few tensions in policy terms between the centre and the locality. It is plausible that local planning officers and their counterparts in central government generally share the same views on appropriate planning policies. In other words, there is no conflict between national and local planning policies because most local policy planners endorse the content of national guidance. If so, the nature of the central-local relations in the development planning process could be said to reflect models of professionalised networks and policy communities, where policy development and implementation responsibilities overlap between the centre and the locality and professionally trained officers share similar values and attitudes at both levels (Rhodes, 1988; Marsh and Rhodes, 1992). However, it is also possible that local planning authorities feel that there is little point in trying to depart from national or regional guidance, given the reserve powers available to central government to enforce compliance.

Without further evidence, it is difficult to come to any firm conclusions here. What is clear, however, is that despite the introduction of a plan-led system, local authorities still operate under tight constraints when preparing their development plans, and deviations from central government's preferred style of policy expression, let alone real conflicts in policy, are often noticed and eradicated through the scrutiny and intervention process. This not only turns local plan-making into something of a painting by numbers exercise but, more importantly, reduces the opportunity for such plans to reflect local differences in policy formulation. In such circumstances, it is not surprising if the resulting plans lack vision and innovation. This is especially true at the strategic (structure plan) level, as recent research sponsored by the DETR has also demonstrated (Baker et al, 1999). More generally, the extent of intervention also contributes to delays in plan-making, further undermining public confidence in the system.

Future Developments and the Role of the Region in the English Development Planning Process

The main theme of this book is that of current issues in European regional planning and development. Readers of this chapter might therefore be entitled to expect that it would have mainly focused on regional land-use planning policy development in the 1990s. Had this been the case it would, however, have been a rather short and negative contribution! The reality is that, throughout the 1990s, the English development planning process has been characterised by an almost complete absence of any real attempts at regional planning. There has been the introduction of regional planning guidance (RPG) for all English regions but the preparation processes involved, and their limited content, reflect the fact that these documents are issued by the Secretary of State. Thus they contain central government policies for the regions, rather than locally determined policies of the region. To date, the centres of power and activity within the English development planning process have thus resided at the centre and the local authority levels and, despite the introduction of the plan-led system, it can also be argued that the local development planning process may actually have became more centrally controlled by the mid-1990s than in the more overtly confrontational era of the 1980s. Thus, far from identifying significant differences in the operation of the plan production process in different areas of England, the survey results presented above largely confirm that central government has both the necessary powers and willingness to intervene in local plan-making processes and that, as a result, there is little variation to be found in the policy content of development plans throughout the surveyed regions.

However, this policy-making and governmental vacuum at the regional level is probably the major reason why the local development planning processes in England has been able to remain so centrally constrained. In the absence of any directly-elected regional tier of government, the main institutional frameworks at the regional level have been the Government Offices for the Regions, which are regional arms of central government departments, and voluntary regional associations of local planning authorities (such as SCEALA in East Anglia) which have attempted to discuss the regional planning agenda and provide inputs to the RPG process. However, the latter have often been hampered by the difficulties of political compromise inherent in such a voluntary, collaborative arrangement. Also they have had to operate with a chronic lack of resources, with most regional planning work being done by working groups of local authority planning officers on top of their normal local planning responsibilities. There has, therefore, been no regional scale of governance or land-use policy to act as an intermediate tier between the centre and the locality.

Since the election of the new Labour administration in May 1997, this situation is beginning to change. Aspects of regionalism have returned to the political agenda, fuelled by pressures emanating from the European Commission as well as developments closer to home in the form of moves towards Scottish and Welsh devolution. The Scottish Parliament and Welsh Assembly were the first initiatives to be legislated for, but these have been closely followed by provisions for a directly-elected Strategic Authority for London and Regional Development Agencies (RDAs) for the eight English regions. The latter, formally established on 1st April 1999, are non-departmental public bodies accountable to Ministers with appointed boards comprising around twelve members drawn from a variety of regional stakeholders, but dominated by business interests. Their aims include the advancement of economic development and regeneration and they will be responsible for preparing an economic strategy for the region.

Although there is still no real evidence of sustained commitment by central government to full regional governance in the English regions, there are also parallel moves to establish voluntary regional chambers which aim to provide a regional voice representing a range of regional interests (DETR, 1997). These emerging chambers are closely linked to existing regional associations (and planning conferences), although at least 30% of their membership must be from non-local authority sectors. Preparatory work in establishing these Chambers is varied across the country but in some areas, such as Yorkshire and Humberside, they have already been launched. Meanwhile, outside the governmental sector, other major organisations and actors, such as infrastructure providers and industry groupings, have caught onto this rapidly developing regional agenda and are actively seeking involvement at this level. Institutional structures within the English regions are thus in a state of considerable flux at the time of writing, and it is as yet too early to accurately predict what will emerge over the next few years.

This time of uncertainty also extends to the future development of regional planning policy. In early 1998, the government issued a consultation paper on the *Future of Regional Planning Guidance* (DETR, 1998) which acknowledged many of the criticisms of the existing RPG series and proposed some significant amendments to enhance and strengthen its future status. These ideas have more recently been incorporated into further draft planning policy guidance (*Draft PPG11*, DETR, 1999). In essence, the intention is that regional associations of local authorities should prepare draft regional planning guidance which will subsequently be tested through a public examination process akin to those currently used in structure planning. A stronger involvement of non-local authority regional stakeholders is also to be expected. The resulting guidance is, however, still to be issued by the Secretary of State, although there is some possibility that this role will be more of a 'rubber-stamping' exercise than has been the situation up to now. The first of the new-style public examinations opened in the Eastern region in February 1999, and work is currently progressing in other regions at a variety of timescales.

This general re-awakening of interest at the regional level and the more specific enhancement of regional planning guidance mechanisms and the establishment of significant new institutional structures (particularly the RDAs) could pave the way for a significant shift of power downwards to the regional level from central government and also, perhaps, partly upwards to the region from the locality as well. In terms of the more specific impacts this might have on central-local relationships in the development planning process, we could perhaps begin to see real progress towards the introduction of a regional level of policy development for the first time since at least the 1970s. There are, however, potential pitfalls. Firstly it remains unclear to what extent the government appointed RDAs will become regional bodies rather than mainly creatures of central government. Secondly, it is uncertain whether the new-style RPGs will be allowed to take on a sufficient regional perspective if they are still to be ultimately issued by the Secretary of State. Thirdly, the relationship between the planning agendas incorporated in RPGs and the economic strategies of the newly-established RDAs is yet to be examined but, if past evidence from the 1960s and 1970s is anything to go by (Baker et al, 1998), there is a serious danger that economic considerations will become dominant, and environmental and other important regional land-use planning considerations could therefore be marginalised.

In the absence of the establishment of a regional tier of government - which could take on over-arching responsibility for both economic development and regional planning, and obtain the democratically-elected mandate to fulfil this role - central government (via the GORs) and local government (via local associations / Chambers) will have a crucial role to play. What will be necessary, however, is enough freedom from both the centre and the locality for the emergence and development of a more regionally-focussed development planning process that is able to conflict with some aspects of national policy and/or result in 'winners' and 'losers' at the local level, if this is necessary for the greater good of the region as a whole. In other words, the English development plan system needs to regain vision, innovation, strategic focus and, perhaps most importantly, become more calibrated to regional needs. As this chapter has demonstrated, this is far from the case at the present – but the current development of regional institutions, governance and policy-making could be the catalyst fore more substantial progress towards these ends. If the opportunity is missed, the high hopes which were placed on the statutory development planning system following the introduction of a plan-led system in the early 1990s will remain unfulfilled and, as a result, the statutory land-use planning system at all levels of government will be the loser.

Note

1   A questionnaire to all local planning authorities in the six regions was used to collect a variety of information about the stages reached in plan-preparation and the extent and nature of DoE intervention during the preparation process. In addition, copies of all written comments and objections made by the DoE in respect of emerging development plans were sought from the local authorities concerned. All comments/objections received were then categorised (according to topic area and source of national / regional policy) and analysed. The survey was undertaken in December 1995 and, following reminders, all responses received by the autumn of 1996 were incorporated into the subsequent analysis. An extremely high questionnaire response rate (around 80%) was achieved, with full information provided by over 60% of the total number of local authorities within the six regions. Further details of the research methodology and the results obtained is published in Baker, 1999.

References

Baker, M. (1998), Planning for the English Regions: a Review of the Secretary of State's Regional Planning Guidance, *Planning Practice & Research*, vol. 13, No. 2, pp. 153-169.

Baker, M. (1999), Intervention or Interference? Central Government Involvement in the Plan-Making Process in Six English Regions, *Town Planning Review*, vol. 70, No. 1, pp. 1-24.

Baker, M., Deas, I. and Wong, C. (1998), Obscure Ritual or Administrative Luxury? Integrating Strategic Planning and Regional Development. In *New Lifestyles, New Regions: Conference Proceedings of the Regional Studies Association Annual Conference, November 1998*, London: Regional Studies Association.

Baker, M., Roberts, P., Lloyd, G. and Williams, G. (1999), *Examination of the Operation and Effectiveness of the Structure Planning Process*, London: DETR.

Brindley, T., Rydin, Y. and Stoker, G. (1989), *Remaking Planning: the Politics of Urban Change in the Thatcher Years*. London: Unwin Hyman.

Cochrane, A. (1993) *Whatever Happened to Local Government*, Buckingham: Open University Press.

Department of the Environment (1992), *PPG1: General Policy and Principles*. London: HMSO.

Department of the Environment (1997), *PPG1: General Policy and Principles*, London: The Stationary Office.

Department of the Environment, Transport & the Regions (1988), *The Future of Regional Planning Guidance: a Consultation Paper*, London: DETR.

Department of the Environment, Transport & the Regions (1999), *PPG11: Regional Planning (Public Consultation Draft)*, London: DETR.

Gatenby, I. and Williams, C. (1992), Section 54a: the Legal and Practical Implications, *Journal of Planning & Environmental Law*, pp. 110-125.

Ho, S.Y. (1997), Scrutiny and Direction: Implications of Government Intervention in the Development Plan Process in England. *Urban Studies*, vol. 34, No. 8, pp. 1259-1274.

Hull, A. and Vigar, G. (1998), Development Plans: Coping with Re-Regulation in the 1990s. In Allmendinger, P. and Thomas, H., *Urban Planning and the New British Right*. London: Routledge.

John, P. (1994), Central-Local Government Relations in the 1980s and 1990s: Towards a Policy Learning Approach, *Local Government Studies*, vol. 20, No. 3, pp. 412-436.

Kitchen, T. (1997), *People, Politics, Policies and Plans: the City Planning Process in Contemporary Britain*. London: Paul Chapman Publishing.

Long, J. (1995), Unaccountable Delay on Route to Plan Adoption, *Planning*, No. 1127, pp. 10.

Loughlin, M. (1986), *Local Government in the Modern State*, London: Sweet & Maxwell.

MacGregor, B. and Ross, A. (1995), Master or Servant: the Changing Role of the Development Plan in the British Planning System, *Town Planning Review*, vol. 66, No. 1, pp. 41-59.

Marsh, D. and Rhodes, R.A.W. (eds.) (1992), *Policy Networks in British Government*, Oxford: Clarendon Press.

Purdue, M. (1994), The Impact of Section 54a, *Journal of Planning and Environmental Law*, May 1994, pp. 399-407.

Quinn, M. (1996), Central government planning policy. In Tewdwr-Jones, M. (ed.) *British Planning Policy in Transition*. London: UCL Press.

Rhodes, R.A.W. (1988), *Beyond Westminster and Whitehall: the Sub-Central Government of Britain*. London: Gower.

Roberts, P. (1996), Regional Planning Guidance in England and Wales: Back to the Future? *Town Planning Review*, vol. 67, No. 1, pp. 97-111.

Roberts, P., Thomas, K. and Williams, G. (eds.) (1999), *Metropolitan Planning in Britain: a Comparative Study*, London: Jessica Kingsley Publishers.

Rosen, B. (1993), Department gets Pedantic about Presumptions, *Planning*, September 1993, pp. 20-21.

Tewdwr-Jones, M. (1994a), The Government's Planning Policy Guidance, *Journal of Planning and Environmental Law*, February 1994, pp. 106-116.

Tewdwr-Jones, M. (1994b), Policy Implications of the Plan-Led Planning System,. *Journal of Planning and Environmental Law*, pp. 584-593.

Thornley, A. (1993), *Urban Planning under Thatcherism: the Challenge of the Market* (2nd ed.). London: Routledge.

Wannop, U. and Cherry, G. (1994), The Development of Regional Planning in the UK, *Planning Perspectives*, No. 9, pp. 29-60.

Young, Sir George. (1992), Keynote address to *Town and Country Planning Summer School*, University of Exeter, 9th September 1992.

9 Co-Regional Planning in Europe: An Idea put into Practice

WILHELM BENFER

Introduction

This paper presents work that is still in progress at the time of the EURRN 1997 conference. The project was started in the spring of 1996 when the *S*tudy of *Co*-regional *P*lanning in *E*urope (SCOPE) project had received all the necessary agreements from the participating partners and funding institutions, including the European Union. It has been based on the assumption that the time has come to break with existing traditions in international comparative regional policy studies and experiment with an approach of co-regional planning in Europe. This idea leaves behind the usual national perspective in such enterprises, in favour of a direct exchange of local and regional experts from different countries. As such, this paper does not try, and cannot, be the product of an academic exercise but rather a report of what has been achieved so far and is intended to provide a stimulus to discuss the underlying ideas. This paper is as much a presentation of the particular approach chosen for the SCOPE project as it is a review of the substantive findings related to the workings of national and EU initiatives in the fields of regional and agricultural policies.

The paper begins by presenting an overview of the SCOPE project, i.e., its objectives and organisation. The second section of the paper will then briefly describe the three regions taking part in the SCOPE project in terms of their regional landscapes, agricultural and regional policies and various development perspectives. The paper will close by presenting recommendations based on the SCOPE project both with regard to the particular approach chosen in this study of co-regional planning and possible changes to the current system of EU policies and programmes in the areas of agriculture and regional development.

The SCOPE Project

Project Objectives

The SCOPE study has basically pursued two objectives. First, it has been designed to provide a forum for the exchange of regional planning experiences and the elaboration of planning perspectives for three rural regions in Europe. Thus, the study's organisational set-up made provision to enable local/regional representatives from the three regions to meet on several occasions. Second, the analyses of the participating regions have been used to evaluate the effects of national, as well as European strategies, in the fields of agricultural and regional policies and to formulate recommendations for their reform. The study was therefore conducted at a time when discussions about the 1999 reforms of EU's structural funds had left the initial stage of administrative contemplation and gained broad-based attention and publicity. In this way, the study's results may be of use in any future elaboration of the European Spatial Planning Perspective. From a 'laboratory situation' the Commission and the member states have been provided with information about the way European regulations actually work in different regions and the potential for inter-regional co-operation within the context of three different planning traditions.

Funding

The SCOPE project is the result of a joint initiative between the Danish Ministry of Energy and Environment *(Miljp- og Energiministeriet)*, the German Federal Research Institute for Regional Geography and Regional Planning *(Bundesforschungsanstalt für Landeskunde und Raumordnung)*, and the Dutch National Spatial Planning Agency *(Rijksplanologische Dienst)*. About half of the total project budget has been jointly funded by these three organisations. The remaining 50 percent of project funds have been made available through a grant from EU's Directorate General XVI, as an Article 10 Pilot Project within the European Regional Development Fund.

Project Organisation

As part of the preparatory work one region had to be selected in each of the three participating countries. In order to ensure a basic level of comparability, the regions had to meet the same criteria, i.e., being intermediary regions in north-west Europe that are, in their respective national context, peripheral, between larger urbanised areas, and facing processes of marginalisation through changes of agricultural land use. By applying these criteria, the selected regions and their development have a strong European dimension, being heavily influenced by a range of subsidies, regulations, and programme initiatives set in the framework of the EU's agricultural and regional policies. Eventually, Ribe county in Jutland, Denmark; the Uckermark in Brandenburg, Germany, and the Oldambt-Westerwolde in Groningen, the Netherlands were selected for the SCOPE project Each area was also willing to contribute a certain amount of manpower towards the operation of the project.

Despite meeting these common criteria these three regions also exhibit signiicant contrasts. Population densities, for instance, range from 40 inhabitants per km^2 in the Uckermark, 71 in Ribe to 167 in the Oldambt/Westerwolde area. Between 1970 and 1994 the population has decreased by 6.2 percent in the Uckermark (with a sharp drop following German re-unification) whereas the Oldambt/Westerwolde area increased its population by 6.4 percent, and Ribe County grew even more spectacularly by 12.1 percent. Although agriculture is relatively important in the rural economy of all the three regions agricultural employment is 5.3 percent in the Oldambt/Westerwolde area, 6.9 percent in Ribe and 9.4 percent in the Uckermark.

Once all preparatory work had been finalised and the necessary project funds been secured, the SCOPE project started its work with the first meeting of the project's working group in March 1996. This group comprised three to four members from each participating country who represented regional and national institutions. The working group was set up to co-ordinate and organise workshops and to document the study's progress, including the writing of the final project report.

In addition to the working group, a similar number of people were recruited from the participating regions to provide particular knowledge inputs. These members were selected on the basis of their familiarity if not personal involvement, in the formulation and implementation of various public policies that determine the development of the region they represented. They also attended the three SCOPE workshops.

A third group also participated in the workshops. These were a small number of local experts who were able to provide detailed knowledge and experience with respect to the specific region where the workshop was being held. They represented particular sectors such as, agriculture, forestry, tourism, economic development etc.

The Dutch National Spatial Planning Agency was the overall project co-ordinator.

Working Process

The workshops were used as a core element of the SCOPE project. During these workshops that were held between June and September 1996, each of the participating regions was given the opportunity to present its main characteristics, developmental problems and strategic policies options. These were then discussed by an international expert panel. During the three workshops, the working methods became more refined. Eventually, a workshop design was developed that employed quite a variety of different methods, including expert presentations, poster sessions, field trips, group sessions, role plays, and scenarios. Together, the workshops proved to be a very effective and efficient mechanism for the exchange of knowledge and experimentation with different planning strategies in a kind of 'laboratory-situation'. Inevitably due to time constraints the workshops could not be expected to produce detailed and thoroughly investigated conclusions and policy recommendations for the future development of the individual regions. However, within their limits the workshops functioned very well as outside review body shedding new light on old problems. These new perspectives were based upon a basic familiarity with the particular kinds of development problems characterising comparable SCOPE regions, but also reflected the different country-specific planning and policy-making traditions that form the minds and souls of the project participants.

The results of the workshops were documented and the general conclusions and recommendations subsequently reviewed by a select group of international experts in the fields of agricultural and regional policies. At the end of the SCOPE project these conclusions and recommendations will be submitted to a panel of national as well as European policy makers.

The Three SCOPE Regions

Regional Landscapes

The landscapes in the three regions are predominatly glacial in character created during the Weichsel and the Saalien glacial period, (i.e. coarse and sandy areas developed under the influence of ice, wind, and melting water). Due to their location adjacent to the North Sea, coastal and marine landscapes can be found in Ribe and Groningen. Sols are the predominant soil type in the most sandy and humid areas of, Ribe and Groningen, and luvisols, gleysols, and podzoluvisols characterise in the Uckermark area. The lightness of many of the soils in the three regions make them sensitive to wind erosion and the low fertility of many of the regions require an extensive use of fertilisers and drainage to secure an adequate yield. Nevertheless in the Dutch context, the polders belong to some of the highest yielding soils in the Netherlands, Uckermark compared to the rest of the State of Brandenburg, has favourable conditions for agricultural use, whereas Ribe, compared to other parts of Denmark, is a low yield region in particular in relation to the production of cereals.

The Ribe is a flat to gently sloping agricultural/woodland dominated landscape. Fields are often enclosed by an extensive network of shelter belts, traditionally one-rowed picea glauca. More recently many of the one-rowed shelter belts have been replaced by three-rowed or broader deciduous shelter belts. The farmstead settlements are mostly derelict, and in areas of very poor soil conditions relics of heathland exist together with small plantations/woodlots. Stream valleys and surrounding low lying areas (mainly concisting arable land or meadows of variable quality) and bigger plantations interrupt the pattern of the shelter belts. Stream valleys, sometimes in combination with the bigger plantations have created newer, less traditional areas. Ninety-five percent of the woodlands are coniferous.

The Uckermark presents itself as a large-scale open agricultural landscape, dotted by bigger woodland areas especially on end-moraine elevations. The ground is undulating and intersected by several stream valleys. In addition, many lakes, bogs, and meadows also add to the remarkable landscape, which is usually not interrupted by buildings. Farms and associated buildings are traditionally situated in the mostly small villages. However, many big industrial farms, formerly run by the state, are scattered throughout the Uckermark. Only few traditional landscape features, such as hedgerows or solitary trees, still exist, mainly as a result of socialist farming practices. The woodlands are mainly for production purposes and are characterised, in most areas, by deciduous species.

The Groningen landscape is characterised by two different types both reflecting its particular genesis. The Oldambt area is a flat, large-scale open landscape dominated by agriculture and like heavy clay soils result in block-shaped fields of considerable size surrounded by drainage canals. Long and narrow strip villages with large farms characterise the landscape. The landscape of Westerwolde is dominated by the River Aa. The edge of the stream valley is lined with old farm land, villages, block shaped plots, lanes and woods. Long, rectangular agricultural plots with scattered farms placed along deciduosly wooded lanes characterise the area.

With the exception of the Oldambt, the landscapes in the participating regions are neither typical nor rare. They are of an average quality and recent intensive agriculture has left its marks to the detriment of both the landscape and ground water quality.

Despite the differences in soil quality agriculture still represents the single most important form of land use in the three regions. Distinct differences can be found in the proportion of land wooded. In Groningen the area is sparsley wooded and this rises to about 20 percent woodland coverage in Uckermark.

The SCOPE Regions within the Framework of EU Policies

All three study regions receive substantial support from the European Union particularly through the Common Agricultural Policy (CAP) and, to a lesser extent, the EU's Regional Policy.

Agricultural Policies. In each of the regions, farmers are supported both by price compensation measures as well as other accompanying measures that were established with the CAP reform of 1992. In terms of inomes, the price compensations that are paid out of the guarantee section of the EAGGF are by far the most important. With the accompanying measures, DG VI (Agriculture) co-finances 25 to 50 percent of the costs in East-Groningen and Ribe, and 75 percent in the Uckermark, due it's Objective 1 status.

With the accompanying measures, the support for environmentally friendly farming in Ribe and the Uckermark differentiates between arable land and permanent grassland. In East-Groningen farmers receive subsidies on environmentally friendly investments irrespective of the kind of agriculture practised. Only in Ribe are some farmers granted assistance for reducing nitrogen use and general grassland upkeep. A small area has been designated for 20 years set-aside, but nobody has yet applied for funding. In Ribe, as well as the Uckermark, farmers receive support for buffer zones. In the Uckermark 25 percent of the agricultural land is designated a less favoured area.

The measures are, generally, an important source of income for the farmers and thus, for the general socio-economic conditions of the regions. In relation to land-use and landscapes the measures are somewhat contradictory. On the one hand, the price compensation measures provide a strong incentive to scale-up agricultural production and the intensity of land-use, resulting in the deterioration of traditional landscape elements, and environmental degradation. On the other hand, and much less significant, the accompanying measures stimulate extensification of agriculture, reduction of pollution, and development of nature. The impacts of these accompanying measures are minor compared to those of the price compensation measures because they represent only a very small element of CAP support.

While the implementation of price compensation payments are handled in a fairly similar manner across the EU member states, the agro-environmental measures are subject to national and regional variations. In the Netherlands, all agro-environmental measures are implemented by LASER, an internally privatised section of the Ministry of Agriculture, Nature Management, and Fisheries. LASER is divided into regional offices, each responsible for the implementation of CAP's accompanying measures. The SCOPE project discovered that LASER is seen as capable of processing the applications without causing friction between the organisation involved. Temporary difficulties only emerge when the rules are changed. Farmers in East-Groningen usually know what kind of information LASER require, and usually enough time to submit their application forms. Moreover, they can get assistance from various regional farmer's organisations.

In Denmark, the measures for organic farming and reduction of nitrogen supply are implemented nationally by the Ministry of Food, Agriculture, and Fisheries. All other agro-environmental measures are implemented in specific areas and by a variety of organisations. In the absence of a comprehensive national strategy, the counties, under guidance of the Agricultural Ministry, are responsible for the designation of areas for environmental farming. This results in considerable differences between the countries in the proportion of land defined as environmentally sensitive. The Ministry of the Environment and Energy is responsible for the afforestation of agricultural land and for the improvement of woodlands and shelter belts. Subsidies are offered to farmers in zones where afforestation is allowed and - with a higher payment rate - in zones where afforestation is desired. For the regionally implemented agro-environmental measures farmers send in their applications to the county council. Within the strict regulatory framework defined by the Agricultural Ministry the county decides whether the farmers will be eligible for support or not. For the farmers it is important that they have the opportunity to select, on a voluntary basis, the measures they consider to be most interesting.

In Germany, some agro-environmental measures are implemented nation-wide, with others it is only parts of certain states. A state only qualifies for assistance from the federal level if its proposed measures fit into the national framework of the 'market and location adapted agriculture' which has been agreed upon jointly by the national and state governments. In addition the states also implement parts of the European agro-environmental programme outside this framework, for instance, with regard to certain kinds of permanent grassland or fallow land. In the Uckermark, with recent application procedures were set up in a complicated and often time-consuming way. Applications had to be sent to the county's agriculture office (*Landwirtschaftsamt*), which forwarded them to two state agencies, i.e., *Amt für Agrarordnung, Landesamt für Ernährung, Landwirtschaft und Flurneuordnung (LELF)*. Finally, it was the state's Ministry of Agriculture which was in charged of giving final approval for most of the measures. At the end of 1996, procedures were simplified putting the *Amt für Agrarordnung* in charge of making the final decision on funding applications. This change has improved the administrative efficiency of the programme and at the same time made it more flexible and able to adapt to local circumstances.

The monitoring and evaluation of the CAP-measures are intense in all the three countries. In Denmark and the Netherlands, the Ministries of Agriculture use several databases, providing information on hectares, crops etc. The measures are also monitored and evaluated by random samples. In the Netherlands, monitoring and evaluation are partly done by satellite surveillance. In Germany, the federal Ministry of Agriculture, the states and the counties are responsible for monitoring and evaluation. This only happens through random sampling. In the three countries, farmers are also controlled by DG VI. Several visits per year by public officials may, however, be considered by some to be too much of control.

Regional Policies In marked contrast the CAP, European regional policy does not have a prominent role in every part of the Union. Within the SCOPE regions, for instance, Ribe County, with the exception of a little island, has not been considered for any of the seven different objectives. East-Groningen, however, is part of an Objective 2 area dominated by problems related to the decline of the potato flour and straw-board industries. The Objective 2 Operational Programme (OP) is dominated by the European Regional Development Fund (ERDF) and largely corresponds with the *Integral Structural Plan.* This plan was produced by the Dutch Economics Ministry together with the northern provinces and was designed to stimulate the economic development of the provinces. The delineation of the programme area in East Groningen, however, is not without its problems. The Objective 2 area has been designated on the basis of problems that are dominant in across the area as a whole, notwithstanding the fact that other problems affect certain parts of the region. Because of this, farmers in East-Groningen only get assistance for the development of tourism. Objective 5b, however, may provide more opportunities for diversifying rural economic activities, and thereby help to counter the threat of marginalisation. Another difficulty is that DG XVI handles the borders of the area very strictly. Businesses submitting project proposals that are interesting for regional development but that are located just outside designated areas are not eligible for project funding.

The Uckermark is an Objective 1 area, a status shared by all of Germany's new Länder. The economic transformation from a centrally planned state economy into a more decentralised social market economy requires enormous effort from public authorities, as well as by private organisations and other groups. Because of the Objective 1 status, the EU co-finances up to 75 percent and in some cases even up to 90 percent of the project costs. The State of Brandenburg has developed two Objective 1 OPs which are dominated by the ERDF and the EAGGF (European Agricultural Guarantee and Guidance Fund). Both OPs are implemented within a national framework. The ERDF OP is formulated within the framework of the 'Joint Task for the Improvement of Regional Economic Structures' (GRW), a 20 year policy initiative of the federal and state governments to bolster regional economic growth. The total amount of funds available for this joint task - which affects all German Länder - is much higher than the part which gets co-financed by the EU. In the Uckermark the total GRW support is six times the support provided by the EU. A similar national policy framework exists for the EAGGF and its OP takes the form of the 'Joint Task for the Improvement of Agricultural Structure and Coastal Protection' (GAK). Whenever an EU regulation is implemented within the GAK, both the federal and the state government share the costs at a ratio of 60:40.

Two of the study areas are supported by the INTERREG II programme. East-Groningen forms part of the Eems-Dollard region, covering the provinces of Groningen and Drenthe in the Netherlands and the border counties of Lower Saxony in Germany. The Uckermark is part of the Euroregion Pomerania, made up of the border counties of the German States of Brandenburg and Mecklenburg-Vorpommern and the Polish district of Szczecin. The INTERREG II OP provides financial support for similar kinds of measures that are supported by the Objective 1 and 2 OPs and also for sport and cultural activities. INTERREG II aims at the integration of local communities on both sides of the border. Therefore, it only supports projects in which partners from both sides co-operate. In the Pomerania region, however, INTERREG II supports only projects on the German side of the border. This is because this Euregion is situated near the EU-border. Projects on the Polish side are supported by PHARE. In the Pomerania euroregion, INTERREG II co-finances a much bigger proportion of project costs compared to the Eems-Dollard region: 75 as opposed to 25 percent. This is due to the Objective 1 status of the new Länder.

As the SCOPE project has shown, several difficulties are associated with the procedures guiding the administration of EU-Regional Policy. The strict procedures for project approval foster an attitude of expectation among potential applicants. This is one of the reasons why in 1995 and 1996 project proposals in the study regions remained below expectations. In the Uckermark, this is enhanced by the strict procedures related to the GRW to which applicants address their requests for support. For the new Länder, however, DG XVI has applied the procedures less strictly because applicants usually lacked experience in dealing with such activities due to their GDR tradition.

The OPs and projects are evaluated and monitored very frequently. Projects are monitored every six months on a ex ante, ongoing and ex post basis. In addition, an audit report is required before the project starts and after its completion. Furthermore, a sample of projects will be subject to evaluation by the European and the national courts of audit. These evaluations require the gathering and processing of great amounts of information, and a lot of complex work, sometimes at the cost of the implementation.

Some criteria related to accessing structural funds are hard to satisfy. This is especially true for the INTERREG II programme in the Uckermark. Some project proposals, e.g. for cultural festivals, have not been approved because the applicants could not prove that their project would not generate economic and social as well as cultural impacts on both sides of the border. Social and cultural impacts, however, can hardly be measured, and it is difficult to generate impacts on both sides of the border if projects are only implemented on the German side.

Sometimes, late approval by DG XVI has proved to be another bottleneck. The approval of the Objective 2 OP for Groningen and the INTERREG II OP for the Pomerania region were delayed for almost one year. For the Objective 2 OP, this was due to a shortage of manpower at DG XVI. For the INTERREG II OP, the main cause of delay was the lack of experience in the new Länder. While DG XVI did allow money that could not be spent in the first programme period to be transferred to the second period, the late approval did cause a delay in the implementation process.

A lack of locally available resources is often a serious drawback for project implementation, particularly in the Uckermar, despite the high proportion of ERDF funding available. In some rare cases, community projects are supported by private initiatives more commonly these project proposals fall by the wayside. As a result, a lack of matching resources causes a shortage of projects to be co-financed by the EU.

Development Perspectives for the SCOPE Regions The workshops held in each of the three SCOPE regions served as a forum at which some rough development perspectives were sketched out. These perspectives were formulated on the basis of an evaluation of the regions' strengths and weaknesses as well as the opportunities and threats. Table 9.1 summarises these preliminary findings and provides the starting point for a discussion of the regions development perspectives. The table highlights the three most important strengths and weaknesses of the study regions.

Despite the many differences that exist between the three SCOPE regions, the development perspectives that have been formulated the SCOPE project do exhibit several common features. For all three regions, agriculture will continue to play an important role in the future and not only in relation to land use patterns. The current structure of farming differs in each region and it probably will continue to do so in the future. All three regions will have to choose how they want to deal with the simultaneous processes of agricultural intensification and a regional economic diversification is absolutely necessary for all regions. Because regional landscape characteristics are cherished in all regions, an offensive landscape planning strategy might be needed.

Of the three regions, Ribe County has the best developed economy and infrastructure. The proximity to the Billund airport is of particular significance as a major development factor. Groningen and the Uckermark

	Most Promising Features	*Most Worrying Features*
RIBE COUNTY	1. Own fodder production in (intensive) livestock farming	1. lack of an integrated regional planning perspective
	2. Development impulses from the Billund urban area	2. unflexibility of rural land-use legislation
	3. landscape renewal by new hedgerows planting	3. absence of integrated Water management
UCKERMARK COUNTY	1. ecological farming and promotion of regional products	1. formalised planning and decision-making culture with lack of flexibility
	2. development potential of the surroundings of the Berlin metropole	2. poor features of agricultural landscapes
	3 landscape quality	3. soil erosion and water management
EAST-GRONINGEN	1. tradition of open and dynamic planning and decision-making	1. low degree of diversification of agriculture
	2. integrated project (blue town) with potential spin-offs for the region	2. poor incentives for small-scale development of forests and nature
	3. potential of a future link between Randstad, Holland, and Hamburg	3. lack of an operational water-system approach

Figure 9.1 Summary Evaluation of the SCOPE Regions

have good access to EU and national funding sources designed to improve their infrastructure and regional economy. Though in the Uckermark it will take a long time for the region to reach an economical and infrastructural level comparable to that of Ribe County.

With respect to agro-tourism a more environmentally friendly or organic agriculture supplying local (organic) food would be helpful to create or keep an attractive environment, increase farm income and tourist attractiveness of all three regions.

To create individual projects and enterprises or to enlarge and diversify existing ones, it is necessary for all three regions to know the internal (local/regional) demand for goods and services. Rural development and (niche) market studies are needed to provide such a knowledge base. Rural culture, images, and characteristics are resources which can be harnessed for the creation of regional goods (e.g. growing hemp for fibres and manufacturing it with a regional logo), food and services. They attract tourists, as well as local consumers, if specific marketing methods are used. Advice and consulting services can be useful to transfer new marketing strategies to local producers.

More intensive agricultural development can also be an opportunity. All three regions share great potential for the further development of industrial milk and meat production, albeit under strict environmental conditions, with the potential of serving world markets.

Development Perspectives for Ribe County Ribe County is characterised by intensive livestock farming (mainly dairy production) with self-production of fodder on mostly irrigated land. Although irrigation of agricultural land is needed to stabilise and increase farm production, groundwater could in the short term be saved by optimising irrigation methods. However, due to the fact that the high quantities of groundwater available for irrigation for agricultural production will be limited in the future, as environmental standards tightened, a change to more extensive farming practices in Ribe County might be necessary in the mid to long term unless intensive industrial agriculture is able to adapt. In any case, more consideration will have to be given to the management of the regional water system as a basis for future development.

On a more positive note, Ribe's human resources are highly educated and this provides good possibilities for the regional economy to adapt to changing living and working conditions.

Landscape management has played an important role in Ribe County in terms of extended plantations and shelter belts, but the diversity of species, especially in the plantation is very low. The Christmas tree plantations might be financially rewarding, but in terms of landscape quality they must be regarded as a threat. As a strategy the region seeks tourists and, there is a need to create more natural attractions and the development of further wetland zones on a regional scale is one idea being pursued. Apart from this, more diversified plantations and shelter belts (e.g. mixed deciduous trees and natural forms of plantations) could help future landscape development.

Development Perspectives for the Uckermark Compared to the other two regions agriculture in the Uckermark is dominated by large farms. Due to the fact that the Uckermark are designated for nature conservation (e.g. the biosphere reserve at Schorfheide-Chorin), some of these farms have applied extensive production practices or belong to organic farming associations. More of such initiatives can be expected in the future, especially as the management for the Biosphere Reserve advocates the idea of organic farming. Projects such as *Women and Environment* help farmers to cope with the considerable variety of extensification measures and may be instrumental in increasing environmentally sensitive land use. However, large-scale industrialised farms will continue to exist in the future despite the efforts to enlarge the proportion of the area that is farmed extensively.

Currently, large-scale farming is dependent on subsidies, given in proportion to farm size. The compensation payments of the 1992 CAP reform will not continue indefinitely making it even more important for farms to diversify and to spread the financial risk between different sources of income, (e.g. farm holidays, non-food production, direct marketing etc.). Farms in the Uckermark have a good chance of obtaining a high level of support for investments in the agriculture sector because of the Objective 1 status.

Whilst there is much potential for exploiting new farming opportunities (e.g. diversify out of food productions into non-food crops to serve the paper industry), or developing new marketing strategies to take advantage of the huge Berlin market (e.g. local demand for organic milk), several mayors in Uckermark County indicated that it is not easy to mobilise people with new ideas or projects. People have faced enough problems to adapting to all the changes associated with re-unification without loosing their own identity. To wait and see may have been a good strategy in the former GDR but it has subsequently lost its value. Rural diversification is not easily developed in a context of low population density, poor regional economy and large-scale production landscape. The umbrella of the Biosphere Reserve may provide an example for a promising strategy in this respect.

Measures are needed to develop people's knowledge try to work against the fact that young, well educated people leave the countryside for more promising alternatives. Much effort has still to be undertaken to improve the living conditions of the local people and to assist in creating or re-discovering rural and village identity which is, apart from creating jobs, necessary for (young) people to stay in the region. An Uckermark project with the twenty village assistants could be a promising. This is a bottom up approach where the village assistants are expected to inform inhabitants about possible development projects, the support opportunities that exist and promote the implementation of such projects.

Tourists visit the Uckermark primarily for its attractive landscape. The Uckermark represents a region with a rich biodiversity and high landscape values. This is reflected in the high percentage of land belonging to the national park *Unteres Odertal* or to the biosphere reserve *Schorfheide-Chorin*. Other areas that have traditionally been used for industrial farming have, however suffered from the steady intensification of land use, with large field sizes being just one indicator. The open and uniform appearance of the production landscape increases the risk of soil erosion and lacks any attractiveness for outdoor recreation. The potential of these parts of the region as a place for living and tourism is rather low. The proximity of the Berlin metropolitan area could, on the other hand, raise the demand for attractive landscapes.

The Uckermark should use the different opportunities the Objective 1 status offers at least until the year 1999 and probably beyond. As before, some municipalities and private investors in the Uckermark are not able to afford their own share of total project costs (between 10 and 25 percent). The development of the structural funds after 1999 should take these kinds of problems into account by allowing a greater flexibility in determining the size of the local share of co-financing and how it can be covered. The application of the subsidiarity principle should take into consideration both the financial situation of regional/local public and private investors as well a the socio-economic and ecological importance of the projects for which financial assistance is sought.

Development Perspectives for East-Groningen In Groningen, efforts to develop new perspectives for agriculture have had little effect. The local agricultural research institute sticks to traditional forms of land use and continues to leave new issues such, as fresh-water management, organic farming, dairy products, landscape management, or forestry, from its research agenda. A horizontal diversification of agriculture and its products, however, could lead to a higher degree of vertical diversification of the agricultural sector. Both sets of strategies would strengthen the region's economic base and enhance the regional attractions for tourists. The landscape would, for instance, with more management, no longer be primarily designed for production purposes. Some old landscape elements could be restored and together with the renewal of traditional farm-building accentuating regional characteristics which could appeal to tourists as well as stengthening local rural identity.

Groningen has a long tradition of large, economically sound arable farms that have over generations neglected the need for change in their working routines. Groningen Farmers from the Groningen area seemed to be slower in adapting to the big changes the last CAP reform introduced. The presence of a rather traditionally orientated agricultural research station in the area may be one example of this. But while people from Groningen have traditionally been employed in agricultural production, high levels of educational attainment and training of the region's workforce offers the potential for East Groningen to restructure its economy. At the same time, the prospects for an intensified industrial milk and meat production appear promising. Such activities should, however, not be undertaken without strict environmental standards being enforced to prevent further agricultural pollution.

In Groningen, the steady intensification and specialisation of farming has resulted in a landscape that is primarily designed for production purposes. Large-scale afforestation projects not only used for production purposes, could improve the region's image as a living and tourist environment. Old traditional farm buildings with traditional architecture lining the area's roads in a bad shape.are in a poor state of disrepair. They could be renovated and used for the accommodation of tourists, residents, or as homes for the elderly. In any case, they would serve as attractions along the county roads.

Regional Attempts to Combine Accompanying CAP-Measures with Structural Funds

This section of the paper on the SCOPE regions should not end without emphasising the potential benefit of a combination of policies and associated funding opportunities as a strategy to increase the effectiveness of public policies for rural development. Several attempts to implement structural funds and accompanying CAP-measures in a combined way have been made in East-Groningen as well as the Uckermark. These initiatives are based on the idea that a combined implementation provides an extra impulse for the development of the region: larger projects can be supported and the resources available can be spent in a more integrated fashion.

The *Blue City Plan* in East-Groningen represents one example of this approach. The plan, which seeks to create an 800 ha lake complex, with 350 ha of nature areas, and the possible afforestation of an area north of Winschoten, has recently been approved by the Province of Groningen. The *Blue City Plan* combines nature development, recreation, and the creation of an attractive environment for dwellings in order to stimulate the development of the region. DG XVI will support that part of the plan that seeks a better utilisation of the landscape and natural environment through tourism. The Minister of Agriculture has negotiated with DG VI about a collective set-aside experiment in order to create the lake complex. This experiment could serve as a first cautious step towards a more regionally attuned CAP. DG VI, however, eventually decided against supporting the plan. The Dutch government has, in the meantime, expressed its determination to provide the support necessary to implement the plan.

An important example in the Uckermark is the village renewal project in Mescherin. Its main objective is to make the village more attractive for residents, entrepreneurs, and visitors. The project includes measures such as the renovation of characteristic agricultural buildings, the (re-)construction of country roads, etc. These activities are to be supported by the GRW and by the EAGGF. The Euro-region Pomerania intends to provide assistance for the renovation of Mescherin's old bastion with INTERREG II funds. Once completed, the bastion which is situated on the banks of the river Oder will also serve as a landing spot for ships thereby stimulating further trade and tourism activities across the German-Polish border.

However, the SCOPE project has indicated that the combined implementation of structural funds and accompanying CAP-measures appears to be not without difficulties. In its effort to prevent the double financing of projects from European sources, the EU has formulated strict regulations guiding the combination of its policies and programmes. Accordingly, combined implementation is only feasible if a programme can be divided up into smaller projects that can be pursued independently. However, by splitting up a coherent project the potential to benefit from synergetic effects are reduced. Moreover, a division requires extra resources for the financing of the separate projects whose co-ordination and administration is made more complicated and time-consuming. Perhaps the most serious disadvantage of such a strategy, the mutual fine-tuning of the individual activities, both in technical as well as financial terms, is made considerably more difficult.

Conclusions and Recommendations

The final section of the paper will deal at some length with the conclusions and recommendations from the SCOPE project. They have been organised around key themes reflecting the two basic objectives of the project, i.e., the development of a specific approach to co-regional studies and the review of EU policies as they unfold in individual regions.

Creating a Forum for Mutual Exchange of Planning Experiences

The workshops showed that regional representatives developed a keen interest in the development problems of other regions and the strategies adopted for overcoming them. This interest may be a reflection of three factors: (a) the study of other regions of the same country may not always provide satisfying answers to the problems experienced at home; (b) the SCOPE project represented an unbureaucratic and flexible format for looking beyond national boundaries; and, (c) the current level of European integration has begun to blur the hitherto sharp contrasts of the individual country's government and regulatory systems that, in the past, seriously limited the co-operation across state borders. With the significance of country-specific traditions in government and public administration continuing to diminish, the potential for regional co-operation across national boundaries will increase.

The SCOPE project has demonstrated that regional co-operation can prove to be very valuable with regard to substantive policy areas either where regions continue to play an important role in the process of policy formulation or where regions are particularly exposed to the implementation of European policies. The latter point is particularly true with regard to rural development. To study rural development in a co-regional fashion has not only offered the possibility of evaluating the effects of individual EU policies and programmes but has also provided insights into the scope and effects of national variations in the implementation of those European initiatives.

SCOPE's focus on regional and agricultural policy revealed that regional planning, despite its different forms, is commonly seen as the appropriate instrument within which conflicts may be mitigated and effective regional development strategies can be co-ordinated. From this perspective, co-regional planning initiatives may well serve as a mechanism to evaluate the combined effects of sectoral policies pursued largely independently from one another by a range of different authorities at the national as well as European levels. This knowledge, if transferred to the respective authorities, could be of vital importance in the process of increasing the effectiveness and further development of European and national policy-making.

Recommendations. The European Union should seek to expand its efforts in encouraging regional co-operation especially in fields where more powers are transferred from national governments to the European level. Establishing good working relations is of vital importance for enhancing mutual understanding between regional authorities, in the field of planning and beyond, across the member states of the European Union.

In light of the looming reform of the European regional and agricultural policies, related issues should continue to be made the subject of regional co-operation. Such activities could provide valuable clues as to the appropriate form and content of future policies aimed at promoting rural development in ways that do not rely on Objective 1 and 5b designations. Promoting rural areas will be of critical importance in the process of increasing the level of cohesion and harmonisation throughout the European Union.

In the broad area of regional policy, regional spatial planning should be considered an issue especially relevant, and of strategic importance for, regional co-operation across national boundaries. Co-regional projects could be used as a means not only for evaluating the aggregate effects of EU policies and programmes but also for pointing out deficiencies in the working of, possibly, the entire range of EU initiatives as they impact upon regional development.

Applying and further elaborating the SCOPE approach

The SCOPE project has demonstrated that the co-operation of regional actors across national boundaries requires outside organisation and moderation. Such assistance was provided by a working group whose members were both knowledgeable in the project's subject matter as well as experienced in project management. This task, (i.e., being responsible for steering the project's progress), is most adequately done by people who participate in the project on the basis of paid contracts.

The knowledge among regional actors about other countries' administrative and regulatory systems tends to be limited. For regional co-operation to be effective considerable effort needs to be made to compile the necessary amount of background information and to make the project participants familiar with country-specific traditions and features of the administrative and regulatory systems as they apply to the individual project regions. One factor proving especially hard to deal with throughout the project has been the availability of comparable data. Differences in the participating countries' systems of data gathering and handling can seriously limit the effectiveness of regional co-operation.

In order to facilitate the participation of regional actors and the exchange of their knowledge and experiences, workshops were held in each of the project regions. The design of the workshops had to be sufficiently flexible to meet with particular regional circumstances. Nevertheless, a set of particular techniques aimed at increasing the effectiveness of the workshops as suitable forms of communication and interaction.

The SCOPE workshops were intended to provide all participants with a sufficient knowledge of the particular local/regional circumstances as the basis to design strategies and individual measures to further the region's development. This objective proved to be appropriate yet at the same time too ambitious given the limited time that was available for the workshops. As a result, the workshop outputs had to be limited in their level of sophistication.

With regard to the participation of regional actors, the SCOPE project revealed that language continues to play an important role in defining the extent and intensity of co-regional exchange of knowledge and experiences across national boundaries. The SCOPE arrangement of choosing a project language that was 'foreign' to all project participants contributed to a fairly balanced level of participation of all the regional actors.

Recommendations. The promotion of inter-regional co-operation by the European Union should not only focus on providing monetary assistance for participating regions directly but also include funding for project management. Such a project management function could be carried out by an arrangement where a group of experts are commissioned to act as project moderators. In light of the increasing importance attached to regional co-operation throughout Europe, such an arrangement could eventually lead to the establishment of several institutions, organised as a network, serving as clearing house, think tank, advisory unit on regional policy etc. with regard to the pursuit of EU regional policy.

As long as co-regional activities undertaken by actors from different countries are still in their infancy, the starting point should be a detailed overview of the regions' government and regulatory systems for the benefit of all project participants. With regard to statistical data, continued efforts should be made to improve the compatibility of the EU member countries' systems of record keeping and data handling.

In order to foster an awareness of the value and benefits of co-regional co-operation, the design of respective projects should consider the need for repeated meetings of the regional representatives whose co-operation is desired. Since meetings of this kind have to be tailored to the specific topics and/or regions under investigation they require considerable resources, careful planning, and the employment of sophisticated techniques for making best use of the, usually very limited, time.

Given the need to involve a sufficient number of local/regional representatives to make co-regional planning work, any such efforts should be capable of providing the means for these representatives to meet frequently and/or for extended periods of time.

In general, the participants of co-regional projects should either all be non-native or all be native speakers. This 'rule' represents a necessary precondition for a balanced involvement of all those taking part in the initiative. If a common foreign language cannot be agreed upon, translation capacities have to be provided to an extent that minimises the differences in language proficiency among the project participants.

Developing Further Regional Co-operation

The interest of the participating local/regional representatives in the SCOPE project was furthered by the expectation that valuable experiences and new perspectives from elsewhere could be used and adopted to fit and perhaps solve some of the problems· at home. However, the amount of time during which a direct exchange of these experiences and perspectives could take place was too limited, tied more or less to the three workshop situations. Nevertheless some regional features that can be of use for others are worthwhile considering:

- Ribe county and the Uckermark could profit from the experiences in East Groningen with an open and integrated planning approach. The regional plan for Groningen and the diversity of planning incentives on the project level could act as a model (not a blueprint!) for further development of regional planning in the other two regions. Mosy notably the recent practice, in the Netherlands, of public involvement in an early stage of programme development could be useful;
- East-Groningen could profit from the experiences with ecological farming (Uckermark) and own fodder production (Ribe County). Looking for possibilities for diversification of agriculture, these features could help the transformation in the Oldambt region to an economically less vulnerable and ecologically more sustainable agriculture;

- Ribe County and East-Groningen could benefit from experiences in the Uckermark with the broad environmentally and socio-economically oriented concept of the Schorfheide-Chorin Biosphere Reserve;
- Ribe offers good insights in establishing a more diversified regional economy;
- The SCOPE project also revealed some commonly shared problems for which solutions should be investigated. Among these problems is the need for developing a water-system-approach as a basis for regional planning. None of the regions has an integrated water management perspective available;
- Finally, the need for an overview of integrated landscape development measures was re-iterated during the project workshops. Although in every region there were some good examples of landscape planning, none of those were able to cope with the dynamics of agriculture, economic and urban environments. Often landscape planning is following, not guiding these developments. The search for a more anticipating, inspiring and flexible form of landscape planning is a common need in all three regions.

Recommendations. To make the best and most effective use of the opportunities that co-regional activities offer for all participants, projects should not be organised on a short-term basis. Provisions should be made that allow for fairly frequent meetings of the participating representatives over a prolonged period of time. This would not only create an opportunity for the exchange of experiences among the participants but also for putting new strategies into practice and jointly discussing and evaluating the results of such actions.

The best way to make crosslinks profitable would be to establish an *Interregional Consulting Programme* (ICP). Experts from one region invite colleagues from other regions to discuss and learn from their plans and projects. Alternatively, experts could go to other regions and investigate the current situation and explore development possibilities. The out-put could be a consulting report and recommendations.

A common lack of experience with respect to integrated water management and landscape development should lead to a co-operative effort toward identifying respective knowledge and expertise elsewhere. Together, the three regions could organise an *Interregional Workshop Programme* (IWP), in which experts from other regions are invited to present and discuss their approach and experience. All three regions could benefit from this exchange in their own ways according their own needs.

Use Spatial Planning for Integration and Regional Direction

Land use in the three SCOPE regions is not only influenced by regional and national policies, but also by DG VI's market regulations on agriculture, set-aside and agro-environmental programmes. Agriculture will at least in terms of land use continue to play a crucial role in shaping wach regions future. In addition, the different EU structural funds can significantly impact upon the process of rural development, particularly in the Uckermark and parts of the Groningen region.

Even though a considerable amount of European money CAP and also Structural Funds isbeing spent in the three regions, this is mainly done without special guidance of a local/regional development plan. Moreover, it is not always possible to make EU subsidies work in a direction which would be considered appropriate for the regions. Activities pursued in the framework of different EU programmes might even counteract each other. The Uckermark, for example, would prefer stronger support of 'sustainable agriculture', which according to the definition applied by local policy-makers should help maintain an open landscape, support environmental goals (e.g. water protection) as well as create and keep employment in the area.

DG XVI provides support from different funds through a plethora of different programmes and measures. In fact, just like the DG VI regulations, their number has risen to a level which is making it difficult to get a overview of the important programme regulations and to design a strategy that makes the best possible use of all the existing EU initiatives for a particular region.

Objectives and specific programmes demand a spatial boundary. Thus, the implementation of useful projects are sometimes limited by the need to correspond with specific programme objectives as well as physical boundaries. Some projects in the Groningen province have had to be divided so that they are soley located within the boundaries of the Objective 2 area.

Co-financing provides a successful stimulus for new, better, and often bigger projects in the regions, furthering the implementation of the additionality principle. On the other hand, co-financing criterion may prevent regional or local authorities from applying for certain EU programme funds because insufficient finances make it impossible for them to provide their share. This has been the case in some Uckermark municipalities.

Experience from the Uckermark has illustrated that the opportunity for combining structural funds can indeed be very useful in achieving integrated development goals. Financial support from EAGGF, ESF and

ERDF has - at least in some way - been combined under a programme for 'integrated rural development' issued by the State of Brandenburg. Nevertheless, even within this regional development framework it was still necessary to divide bigger projects into smaller parts, with each component having to apply for money from a variety of sources, essential on their own. Moreover, the co-ordination mechanisms and agreements between respective authorities, required by such an approach, do not always function adequately.

The implementation of detailed and complex programmes often requires much time. Operational programmes for Objective 2 in Groningen and the INTERREG II programme in the Pomerania Euro-region, of which the Uckermark is a part, were finally approved by DG XVI almost one year after the beginning of the planning perithese programmes should have begun.

The existing way of spending most of the DG VI funds in the study regions, i.e., in form of a considerable number of different, uncoordinated individual measures for market regulation in a rigid pan-European framework, does not necessarily take into account the specific regional needs. With the exception of some accompanying measures, this funding practice does not effectively cope with new demands, such as, the need for sustainable agriculture or environmentally sensitive farming or the need to satsify social criteria. In addition, the multitude of different programmes creates confusion for farmers as to which scheme is the best to pursue. This can be critical where the optimum use of such programmes is vital to securing the economic base of the agricultural sector and, hence, the region as a whole.

The highest proportion of DG VI funds spent on agriculture is directed towards price compensation payments, with no direct linkage to environmental, social, or landscapes management criteria. Because of the absence of these linkages, compensation payments may be working against existing market forces and public opinion that are becoming more sympathetic towards the need to apply social and ecological criteria to agricultural support mechanisms.

Subsidies are frequently given for initiatives that further aggravate a region's ecological problems, instead of solving them. This situation prevails even though some new incentives have been introduced in recent years, e.g. the Regulation 2078/92 aims at ameliorating the negative impacts of intensive agricultural production. Such as agro-environmental programmes, are sometimes not well accepted by arable farmers, especially those with arable lands. However, the funding levels of these programmes do usually not match the amounts of money available through price compensation payments. Thus, a farmer in the Uckermark may get twice the amount of subsidies for growing rape than for special water protection measures. In addition there are specific incompatibility problems with regard to existing agricultural management practices, e.g., extensive grass production may not fulfil the fodder quantity nor quality needs for milking cows.

Recommendations. Specific regional conditions and development problems demand a regionally specific approach in accordance with existing regional development goals. In this regard, spatial planning at the regional level could function as an appropriate instrument for formulating and implementing these regional development goals.

Successful development of rural areas could be further supported if CAP and Structural Fund measures, could be strategically used in a more flexible and regionally specific way within a coherent, broad framework of European Union policy guidelines. A spatial planning perspective could contribute to this difficult challenge.

Moreover, for such a strategy to work, regional development plans should pay specific attention to those measures eligible for EU subsidies and at the same time the reform of the EU subsidy system should consider a closer connection with regional spatial planning. This would open up the possibility of designing comprehensive financial support schemes that respond more adequately to the local and regional needs. Eventually, the regions could be able to apply for subsidies for integrated projects, which are locally formulated according to specific needs, current practice where programmes are designed to tackle local problems in ways that best fit the tight and detailed rules set up to administer EU support programmes.

Therefore, it would be useful to create 'regional development platforms', where discussions and negotiations between the different funding authorities, together with regional authorities and non-government organisations could take place. As one of their tasks, these 'platforms' should make attempts to pool financial resources available through different funds at the regional level in order to support integrated development goals as well as large-scale. The Blue Town Plan in Groningen offers a possible model.

Flexibility of DG XVI measures will increase especially if the implementation of objectives and programmes is not tied to specific spatial/administrative boundaries, unless strictly necessary. Where needed, these boundaries should not demarcate lines of exclusion from eligibility but rather differentiate between degrees of eligibility.

In principle, the system of co-financing represents an appropriate approach to manage the funding of regional development activities.

206 Regional Planning and Development in Europe

Nevertheless, the current system of fixed percentages of EU-funding should be made more flexible by relating to the relative wealth of a region as measured, for instance, by the gross domestic product per capita or in terms of the human development index. In addition, flexibility in funding should also be applicable to the now strict precondition of national co-financing.

Further reform of the CAP should lead to a shift of funding priorities away from market regulation towards integrated rural development approaches. Aspects such as environmental, social, and landscape quality should be integrated or added to the list of eligibility criteria for EU subsidies. This will support an integrated approach to regional development. If landscape quality are used as a criterion for EU spending it would also be an investment in the sustainability of the environment and nature as well aspromoting the rural economy. Funding for existing schemes, which already take into account such criteria, e.g., Regulation 2078/92, could be expanded and new models for rural development support should be investigated (such as models for sustainable agriculture, delivery mechanisms for rewarding farmers for production of landscape or biodiversity etc.). New approaches of this kind should also be implemented within the framework of a corresponding spatial planning perspective.

Thus the co-regional approach to rural development planning as epitomised by the SCOPE project provides the opportunity for collaborative learning and also suggest ways that EU support, particularly in relation to CAP and Structural Fund reform could lead to more effective integrated rural development. the extent to which these lessons have to be learnt is something only time will tell.

Acknowledgement

This chapter is based on the work of the SCOPE Working Group and its members Gisela Beckmann, Wilhelm Benfer, Herman Haverkort, Lone Kristensen, Doris Pick, Henrik Praestholm, Jorgen Primdahl, Peter Smeets, Diethard Städtler, and Marcel Wijermans.

10 The Local not the Global? Regulating Local Environmental Policy Making Regimes

DAVID GIBBS AND ANDREW JONAS

Introduction

Recent national and international environmental policy initiatives have placed considerable emphasis upon the local level as the most appropriate site for policy intervention. For example, policy statements on the implementation and delivery of sustainable development such as the United Nations Agenda 21 Programme (through Local Agenda 21), and the European Union's Fifth Environmental Action Programme 'Towards Sustainability', place particular stress upon the local scale for policy implementation. In the United States, local-level approaches to balancing habitat conservation and development have recently been endorsed by the federal government, including the State of California's Natural Community Conservation Plan (NCCP) initiative which promises to protect valuable ecosystems in Southern California from further destruction by urban development. On an international scale, then, there is a discernible move towards local solutions to seemingly intractable conflicts between environmental protection and economic development.

Although there has been a move at the international and national levels to decentralize environmental policy making, the rationale for this move has often been couched in vague terms, such as a felt need to incorporate local stakeholders into decision-making structures or reference to the division of regulatory powers and responsibilities within the state. For example, in the U.S. context federalism and the relative autonomy of local levels of government strengthen the case for decentralization. In the U.K., however, decentralization could be interpreted as another example of central government passing responsibility down to more local levels. But even in this context changes have been occurring in the governance of local economies which suggest there is a case for including local interests and groups into the policy making process (see Cochrane, 1993), especially given that traditional centrist or 'command and control' measures on the environment have clearly failed to generate long-term approaches to reconciling environmental protection and economic development.

So much for the policy rationale for the local scale. What about the academic case for a focus on local interests and groups in the environmental policy making process? How does local environmental policy making intersect with more general processes relating to the governance and regulation of local economies? It is fair to say that the environmental policy community has focused its attention mainly on developments at the international and national levels. This is perhaps understandable in light of the global importance of environmental issues such as greenhouse gases and the loss of biodiversity on a planetary scale. It also makes sense in light of the increasing globalization of economic activity which has posed immense problems for the regulation of environmental impacts by national and local governments. But it begs the question as to how to approach the local environmental policy context. Should the local be seen as subordinate to a wider - national and international - set of economic and regulatory processes and dilemmas? Or does the local have some degree of causal autonomy such that it is possible to identify and conceptualize local interests and groups as constituting relatively coherent local policy or political 'regimes'?

We believe that there are promising avenues for exploring at both a relatively abstract level, as well as more concretely, the links between questions of local governance and issues of local environmental policy making. Indeed, we suggest that, viewed as a *process* local environmental policy making cannot be analyzed separately from that of the governance and regulation of local economies. Yet while researchers have recently begun to approach the latter using a 'reconstructed urban regime theory' approach (Lauria, 1997), the incorporation of local environmental policy making into concepts of regime and social regulation has not yet occurred. This is despite some promising moves in that direction, such as the inclusion of local conservation issues and conflicts into the analysis of local political regimes (e.g. Jonas, 1997), and the adoption of transaction cost approaches to the analysis of local environmental policy making (e.g. Angel, Jonas and Theyel, 1995).

In this chapter, we approach the process of local environmental policy making from the perspective of urban regime theory (see Stone, 1989; 1993) but also with a view to the wider context of regulation approaches (see Lauria, 1997). We are particularly concerned with the following questions: Is it possible to identify relatively coherent 'local environmental policy regimes'? If so, how are such regimes inserted into, shaped by, and in turn shape, wider regulatory processes and structures? We suggest that the answers to these questions are largely empirical, while the questions themselves are motivated by a desire to address the local dynamics of accumulation and social reproduction within capitalism.

Towards an Approach to Local Environmental Policy Making Regimes

It is clear that we need to move beyond a view which seeks to divorce the local environmental policy making process from the broader issue of the governance and regulation of local economies. The danger of such a view is that it treats the 'environment' as a relatively self-contained and closed system, the constituent elements of which can be monitored, modeled and, subsequently, regulated with little regard to its interactions with 'external' economic and political systems. Although there is a growing recognition of the need to develop an integrated and holistic approach to the environment and economic development at the local level through local authority action and the concept of the 'sustainable city' (Gibbs, 1996; 1997). In practice such an approach has failed, not for want of policies so much as the failure of those policies to take into account the constituent properties of local economic and political systems. The fact that local environmental conflicts continue to occur around problems of externality (e.g. NIMBYism) or a failure to internalize transaction costs associated with environmental regulation suggests to us that the relationship between governance and environmental sustainability at local level is far more complex than previously imagined.

It may be that, in terms of analyzing the policy making process, particular problems occur when those relationships are projected onto a spatial context. Traditionally, the relationship between environmental policy and space has been viewed in one of two ways, as an externality or as a cost. Firstly, the spatial context can be incorporated as an 'externality effect'. Thus one rationale for decentralizing environmental policy would assume that environmental impacts either already are, or in the future can be, confined within well-defined local or regional physical systems, such as ecosystems, watersheds and catchment areas. The trick in terms of policy then is to determine the boundaries of these systems and attempt to introduce policies and regulations consistent with those boundaries. Such policies and regulations would aim to internalize externalities arising from, for example, property and access rights, overlapping jurisdictional authority, and so forth. The failure to internalize externalities may cause conflicts which, in turn, highlight the problem of coordinating the territorial division of labour. Solutions to these conflicts usually involve higher tiers of government or the organization of new levels of jurisdictional authority (Cox and Mair, 1991). In this context, conflicts around the environment are not so much eliminated as displaced territorially.

The second way of incorporating space into environmental policy making is to treat it like any other 'factor of production' - as a cost, either to business or to the consumer. Thus any business owner, consumer or property owner who by virtue of their location is impacted by environmental policies or regulation must factor in 'the environment' as a potential cost (or benefit) depending on the nature of the policy or regulation. In this context, the effectiveness of local environmental policy making can be gauged by the extent to which local stakeholders are incorporated into the policy making process and full account is taken of their actual, or potential, transaction costs (for example, using contingent valuation methods). Transaction cost approaches to environmental policy making suggest that that policy consensus is more likely to occur when participants have built up trust based on the frequency of local contacts or have long-standing interests in the locality (Angel, Jonas and Theyel, 1994). Conversely, a failure to negotiate and internalize transaction costs can be a measure of the level of local resistance to a particular set of environmental policies or regulations. In the United States, for example, the failure to compensate property owners adequately for development restrictions resulting from local conservation measures has encouraged litigation all the way to the federal Supreme Court.

Problems of incorporating space into an analysis of the environmental policy making process highlight an issue generic to the governance of local economies in capitalism, that is, the problem of managing the (spatial) division of labour. This problem has been addressed by urban regime theory, which for the most part treats it as an organizational issue (Elkin, 1987). Accordingly, conflicts arising from the division of labour between state and market actors may be contained by governing coalitions made up of whichever interest groups are the dominant or hegemonic forces in the locality (Stone, 1989). Since any given locality will over the course of time have developed distinctive economic and political institutions, local political regimes are historically contingent (Ramsay, 1996). The nature of the local political regime can change, as can its policies:

> Urban regime theory asks how and under what conditions do different types of governing coalitions emerge, consolidate, and become hegemonic or devolve and transform (Lauria, 1997, 1-2).

Urban regime concepts can therefore usefully be deployed to analyze the characteristics of local governing coalitions and their policies. In the case of local environmental policy regimes, this would focus on interactions between firms, local politicians, environmental enforcement agencies (both local and national) and environmental pressure groups, as well as a whole host of activities that may not immediately be thought of as having 'environmental' consequences. It might also be possible to examine specific institutional developments within a locality resulting from attempts to internalize externalities and transaction costs relating to specific environmental policies and regulations. Conversely, a failure to internalize such costs and externalities locally would provide some indication of the 'boundaries' of the local environmental policy regime, and the extent to which non-local powers and capacities come to bear upon the local policy process.

However, regime theory may not be able to say much about the longer-term stability and coherence of local environmental policy regimes. Indeed, one of the major weaknesses of urban regime theory lies its failure to link the structure of, as well as changes in, the local political regime to wider systems of accumulation, modes of regulation, and state policy shifts (Lauria, 1997). It is especially weak on the spatial context, notably the impact of changing spatial divisions of labour on the form of governing coalitions and territorial structures in the state. As Cox (1997) suggests, regime theory has failed to theorize in an adequate fashion fundamental interests in local economic development and the contingent conditions - policies, regulations, state structures, etc. - under which those interests are realized locally.

To address this weakness we suggest that a regulationist perspective on local political regimes allows for a more contextualized analysis of

> the strategic capacities rooted in local institutional structure and organizations (Jessop, 1997, 63).

In this respect, 'the environment' is best viewed as both an object of regulation and a focus of struggle between local actors and groups, each of which may incur different costs (or benefits) as a result of the types of policies and regulations introduced. Indeed, a reconstructed regime theory incorporating a regulationist perspective can be developed through understanding the processes and struggles involved in the formation, reproduction and crisis conditions of local political regimes and wider modes of regulation. Local environmental policy regimes are dynamic forms that are in a continual process of formation and becoming, in the face of strong challenges and countervailing pressures (Painter, 1997).

We need, therefore, to examine through empirical research the ways in which the environment is utilized, interpreted and fought over rather than making the assumption that delivering environmental improvement and sustainable development is a straightforward function at the local level. We now turn to case studies of local environmental policy making in Great Britain and the United States to illustrate the kind of approach we have in mind.

Environmental Policy Making in England and Wales: Local Economic Development and the Environment

Within England and Wales there has been an increasing interest in the local as an arena of environmental policy intervention. The focus on this scale has come from two directions. First, from the 'top-down' policy developments at international (Agenda 21), supra-national (the European Union's Fifth Environmental Action Programme 'Towards Sustainability') and national scales (the Government White Paper, *'This Common Inheritance'*, the UK sustainable development strategy and its revisions to planning guidance notes) which place a particular emphasis upon the local as the key site for policy implementation and delivery. Second, there have been a number of 'bottom-up' responses, largely engendered by local authorities and galvanized by the Local Agenda 21 process, suggesting that the local authorities themselves have attempted to engage in discussions with local interests and groups potentially affected by environmental regulations. As a consequence, many local authorities have devised, and begun to implement, their own environmental strategies to take greater account of the environment within the broad set of their own activities. In this section we draw upon recent research conducted in England and Wales which focused upon the extent to which environmental policy was being integrated with local economic development policy at the local level (Gibbs et al., 1996, 1998). A critical issue in this study was the analysis of strategy and discourse in the local environmental policy making process.

This case study work conducted with local authorities in England and Wales revealed a diversity of opinion and uses of 'the environment' within local authorities and its incorporation into the local policy making process. The case study local authorities had all developed an overarching environmental strategy which was intended to inform and shape other local policy areas. Such strategies had usually emanated from departments with an environmental remit (usually a separate environment department or located in planning). The objective of these strategies was to encourage an holistic interpretation of the environment across all local policy areas, incorporating a variety of local stakeholders. Such local environmental policies and initiatives are obviously intended to have local impacts, such as local environmental improvement, but they are also intended to make a contribution to the resolution of global environmental issues. In broad strategic terms then, the 'environment' is seen as something beyond the local.

However, the ways in which environmental policy making engaged with the key area of local economic development policy revealed the adoption of a discourse which allows the incorporation of 'the environment' into existing local regimes in a non-threatening manner. For example, one form of engagement with the 'environment' for economic development departments comes through the need to take account of European policy. In order to obtain European funding, local authorities need to take account of the EU's regulatory requirements. For example, the Community Support Frameworks (CSFs) which form the contracts between Member States and the Commission require conformity with environmental legislation and an assessment of the environmental impact of projects. The revised Structural Fund Regulations adopted in 1993 require Member States seeking Structural Fund support to produce environmental profiles of the regions in their Regional Development Plans. The case study local authorities were increasingly impelled by these regulations to take account of the environment within their economic development proposals. However, incorporating environmental issues had largely occurred through invoking infrastructural improvements to the local physical environment that were intended to support existing pro-growth approaches to economic development within the local area, such as inward investment and the development of prestige projects. The 'environment' in these cases is about the 'image of the area', such as landscaping, tree planting and the creation of an 'attractive environment' for potential investors, closely tied in with other regeneration attempts (Gibbs et al., 1998).

This appropriation links closely with Hay and Jessop's (1995, 9) concepts of 'discursive mirroring' and 'discursive/policy transfer' whereby

> resources are increasingly distributed in terms of contenders' ability to appropriate a particular discursive style and give it a local inflexion.

In this case, however, the particular local inflexion to the discursive style largely negates the aims of local environmental policy making. In large part this is a consequence of local policy makers having to address two incompatible discourses emanating from both EU and national levels. One discourse revolves around freeing markets, creating greater flexibility and, until recently, a neo-liberal agenda, whereas the other discourse revolves around notions of sustainable development and greater social equity, with the implied need for a higher degree of intervention. Given the incompatibility of these two at any spatial scale, it is perhaps not surprising that local authorities find it difficult to reconcile the two at the *local* level.

While some commentators/academics have been optimistic about the translation of a growing concern for local environmental policy into strong action at the local level, we would agree with Healey and Shaw (1994, 426) who, drawing on the evidence from past planning history, argue that

> past history is not reassuring about the potential for increased leverage of environmental considerations over those of economic development. The new agenda of environmental issues may continue to be drowned out by a dominant materialism.

The case study work with local authorities in England and Wales tends to support this more sceptical approach. Concern for environmental issues within local economic policy making, let alone sustainable development, is largely confined to environmental and planning departments within urban local authorities in the UK. In cases where environmental issues have been addressed in economic development strategies, these have been translated into forms that either do not threaten the status quo, such as the clean up of contaminated land, or are seen as directly complementary to entrepreneurial activities, such as improvement of the physical environment to encourage inward investment. In these cases the environment has essentially become a commodity to be repackaged for corporate consumption. The 'entrepreneurial local authority' thus emphasizes inward investment, property development and job creation over environmental targets and the result of constructing new buildings and roads will be increased use of energy, transport and materials. In the case of some central government policy initiatives, such as City Challenge and the Single Regeneration Budget (SRB), these attitudes are given institutional force through viewing the environmental aims of economic development as targets for land reclamation, improvement and kilometers of roads built or improved. Indeed, the objectives embodied in initiatives such as SRB, which are very influential in shaping economic development schemes, act as a severe disincentive to incorporating environmental objectives (Gibbs, 1997).

Although the conclusions at this stage must be tentative (and we are wary of utilizing the case study information for this purpose) the case studies tend to indicate that the dominant (hegemonic) regime at the local level is pro-development, reflecting the importance of growth strategies within local economic development. Although the environment is very much on the policy agenda of local authorities, an analysis of discursive practices suggests that it has not displaced local economic development as the principal local authority agenda. What evidence there is for an integrative approach to environmental policy and economic development suggests that discursive practices are being developed which are consistent with pro-growth local policies.

Admittedly, this work concentrated solely upon local authorities and so may not have identified relatively self-contained local environmental policy networks. However, it did reveal the involvement of groups external to the local authority in policy development and initiatives, indicating that a reconstructed urban regime approach holds possibilities of identifying a wider range of local interests and groups. However, it is notable that in the U.K. context the power relationships within local authorities reflect the dominance of more traditional economic development forces, and the relative weakness of planning and environmental departments.

From this evidence, the local environment in England and Wales as an object of regulation and as a political struggle between groups/local actors is being used as a support for pro-growth economic development. In some ways this is surprising given the growth of Local Agenda 21 within the UK and the seeming importance of the environment within local policy making. This might be explained, in part, by the non-involvement of the private sector in most Local Agenda 21 strategies (see for example Newby and Bell (1996) on the failure of Leicester's Environment City initiative to involve business and developers in the process). In this sense, then, the 'local environmental policy regime' omits one of its most important (and contentious) elements. Plans and policies on the local environment being formulated within UK local authorities are naive about the wider regulatory structures within which they are situated. The 'environment lobby' within local areas has perhaps overestimated the potential for change and their ability to make a difference in the context of a regulatory structure which privileges development at the 'expense' of the environment.

Environmental Policy Making in Southern California: Habitat Conservation Planning

A major environmental issue in Southern California is the conservation and protection of habitat for species threatened or endangered by urban development. Although the federal government has a well-recognized set of laws to protect threatened species contained in the 1973 Endangered Species Act (ESA) and subsequent amendments, the enforcement of the ESA has proven problematic particularly when it has involved protection of threatened habitat on private property. However, recent local initiatives in California as well as in other States suggests that ways are being found of incorporating local interests and groups into the habitat conservation planning process. One of the most recent and publicized of these initiatives has been California's Natural Community Conservation Plan (NCCP).

To understand how the NCCP framework has developed it is important to consider the prevailing system of land use planning and development in Southern California. The region has in less than a century been transformed into an urban-industrial landscape of global significance. In the process, natural and agricultural landscapes including riparian corridors, coastal sagescrub, chaparral and citrus groves have been all but destroyed, and with these the habitat of numerous species. Much of the destruction of open space and habitats has in recent years been caused by rapid urban development and, in particular, suburbanization: the spread and encroachment of residential subdivisions on open space in outer metropolitan areas. Such development has put increasing pressures on the remaining open space, thereby contributing to a conflict between private development rights and public interest in the protection of open space.

In California, land use planning - hence the regulation of open space on private property - is first and foremost a local authority function. Despite historic attempts to shift land use planning and regulation upwards, including the so-called Quiet Revolution which at one time promised to shift control to the state level, it is municipal and county governments which establish zoning guidelines and subject development plans to public review and scrutiny. Yet because local government has traditionally been a forum for pro-growth groups - the property development coalitions identified by Logan and Molotch (1987) - to exert influence over land use planning, it has become increasingly difficult for environmentalists and local residents to use existing local land use regulations to preserve and protect open space. In Southern California, this situation is complicated by the so-called 'fiscalization of land use' whereby local governments have become increasingly dependent on local revenues from the sale and transfer of property and development. In short, local government in Southern California has a built-in interest in supporting land development activity rather than the preservation of open space. In this context, the process of local environmental policy making takes place in an institutional framework of well-organized policy regimes operating at the municipal and county levels.

Local interests and groups have mobilized in response to interim land use controls imposed under the terms of the ESA. Passed in 1973, the ESA prevents any harm being done to, or take of, species listed as endangered. Habitat conservation measures under the Federal ESA are enforced by the US Fish and Wildlife Service (USFWS). Under an agreement reached in 1982, the USFWS allows landowners and developers some 'take' (or destruction) of species so long as they enter into an agreement with a local conservation agency or the USFWS such that actions are taken to protect the habitat of federally-listed species. Such habitat conservation plans (HCPs) are designed to bring together stakeholder groups whose interests are affected by the listing of a particular species or groups of species. Well over 350 HCPs are currently being negotiated across the USA, and a number involve counties and municipalities in Southern California.

Incorporating local interests and groups into the HCP process has not proven easy, especially in Southern California. Environmentalists have complained that HCPs amount to little more than 'habitat development agreements' which allow developers and property owners to destroy habitat with little consideration for the long term and changing needs of the species they are designed to protect. HCPs have also been unpopular with developers, both because they involve interim land use controls and because they are under funded by the federal government. Some smallholders have felt that their interests are not adequately represented because local authorities involved in developing HCPs seek advice from the housing industry rather than from local farmers and property owners. Moreover, property owners subjected to conservation-related restrictions on development have not always received adequate financial compensation.

In Southern California, local initiatives to protect open space and private property from conservation-related development restrictions have resulted in the emergence of three types of local environmental policy making regime:[1]

- Pro-growth regime: In Riverside County, where landownership is fragmented, the organized housing industry has heavily influenced development of local habitat conservation plans. However, the interests of smallholders and environmentalists have been all but excluded from the conservation planning process. Conservation measures have been restricted mainly to property already under public ownership.
- Entrepreneurial regime: In Orange County, landownership is consolidated. A single landowner, the Irvine Ranch Company, has been pro-active in seeking conservation measures which consolidate long term development rights. Amongst other actions, this has led to sub-regional planning under the NCCP (see below).
- Growth management regime: In San Diego County, the development of local multi-species conservation plans has been driven by the local municipalities, with a leading role played by the City of San Diego. San Diego has combined conservation measures with a long-term plan to expand its wastewater treatment system and limit urban development. This plan creates, in effect, an Urban Limit Line for the City of San Diego and has found widespread support from citizens' groups, regional utilities, developer organizations and conservationists.

In the context of Southern California, then, it is possible to understand the trajectory of local environmental policy making in the context of these three quite distinctive policy regimes. These regimes have developed their distinctive features from the history of development and politics in each of the three counties. But there also is the sense that they share a common trajectory which unites their local interests to the wider mode of social regulation in Southern California.

As Peck and Tickell (1992) suggest, one feature of uneven development within national economies is the tendency for regions to develop their own 'local modes of social regulation'. Elsewhere (see Jonas, 1997), one of us has argued that since at least World War Two the prevailing/hegemonic mode of social regulation in Southern California has linked that region's economic fortunes to the wider 'suburban-defense economy'. It appears, however, that this mode of social regulation is in a moment of crisis and transition. In this context, local interests and groups have mobilized through local political regimes to introduce policies, etc. which have led to a revival of the region's economy. Local land use mechanisms such as habitat conservation plans, development agreements, the transfer of development rights and redevelopment have been developed in ways which encourage further rounds of inward investment and land development (Jonas, 1997).

In this rapidly changing wider context, the stability of existing policy regimes must remain in some doubt, and hence also the conservation measures which they have introduced. It is clear, for example, that some local interests and groups have been turning to non-local levels to circumvent the local environmental policy making process. In Riverside County, uncompensated property has sought redress through the courts. Recent federal Supreme Court decisions have upheld the opinion that private property taken for conservation purposes must be duly compensated and, moreover, any individual or group has the right to file suit against a local or federal conservation agency. Other groups, including the housing industry, have pushed for changes in the federal Endangered Species Act and for clarification of laws relating to the take or harm of species.

In San Diego, the conflict is predominantly territorial in nature - a conflict, that is, between the City and County of San Diego over future development in unincorporated areas. This conflict is likely to feature in the debates around the funding of San Diego's multi-species plan, particularly whether or not the city will seek the passage of a bond issue for its plan at the county rather than the municipal level.

However, there have also been moves locally to forestall these sorts of conflict by introducing a regional or sub-regional component into local conservation initiatives. The most recent of these, and arguably the most significant in terms of developing an integrated approach to conservation and development, is the Natural Community Conservation Plan (NCCP), which was introduced by the California legislature in 1991 in response to pressures from landowners, developers and conservationists in Southern California. In contrast to the single species or local area approach of HCPs, the NCCP approach emphasizes conservation planning on a multi-species and ecosystem (sub-regional) planning basis. It incorporates existing local agreements between landowners, local authorities and conservationists regarding the setting aside of threatened habitat and the implementation of long-term management plans. If pilot schemes in San Diego and Orange counties prove successful, the NCCP promises to establish limits on urban development and habitat fragmentation in Southern California. In this respect, it marks the first, if tentative, step towards an integrative approach to conservation and development in one of the world's most dynamic economic regions.

Conclusions: Towards a more Coherent Approach to Analyzing Local Environmental Policy Making?

Our argument about developing a new approach to the analysis of local environmental policy making is driven by perceived shortcomings in the environmental policy literature as well as wider developments which have highlighted the local scale as a terrain of policy intervention. We have so far in our discussion deliberately not addressed the issue of sustainability. For us, this begs the issue of what is being sustained, an issue which cannot be addressed in abstraction from particular local interests and groups. In other words, for us the issue of sustainable development cannot be treated separately from the sustainability (coherence) of local environmental policy making regimes.

There is a case to be made that the governance of local economies does in the longer term attain some level of self-regulation and can even appear to be coherent and stable. This, at least, is a tendency in capitalism which has been suggested by writers like David Harvey (Harvey, 1985). It suggests to us that the local scale is not only an appropriate point of analysis into the environmental policy making process, it is also a useful point of intervention. It offers a terrain whereupon the interest groups and their policy networks can become fairly coherent and hence identifiable as relatively self-contained - if not in actuality - closed systems. Certainly, the identification of the principal interests and groups, and their arenas of interaction, discourses, and substantive policies, goes a long way towards identifying the 'boundaries' of a local environmental policy making system, albeit we would be hesitant to suggest that these will correspond to already existing physical, property or even local jurisdictional boundaries.

While regime theory can tell us a lot about the process of local environmental policy making, we cannot say for sure how enduring (sustainable?) resultant local environmental policies are likely to be. But that said, if the current capacity to effect meaningful change and positive action on the environment at the local scale is limited by globalization tendencies and nation-state hollowing out, it will be even more limited insofar as it fails to address, let alone challenge, the dominant power relations and politics of particular localities. As Lauria (1997, 7) comments we can hypothesize that the structure of capital and representation of capital fractions locally will affect the nature of the governing coalition, potential urban regimes and their development strategies.

Without a way of grasping the internal dynamics of local environmental policy making regime, it would be difficult - even impossible - to identify problems of externality and transaction cost leakage, so often sources of conflict around the environment and economic development.

So we provisionally reject the notion that it is impossible to approach the issue of sustainability at the local level without understanding hegemonic forces arguably operating at a more global level. Rather the implementation of sustainable environmental policies demands understanding and action at different, yet strategic, points and scales of intervention, some of which are likely to be more local than others. For us, a reconstructed regime approach to environmental policy making offers a useful starting point. It provides more substance to the claim in policy making circles that the local scale is the appropriate level at which to deliver sustainable policies on the environment. This reconstructed approach would combine regime theory with regulationist approaches to offer a theoretical context which can help to inform us of the processes at work. From this perspective, modes of social regulation have a variety of forms and levels of abstraction from the 'real regulation' of laws and concrete structures through to more intangible elements, such as values and norms of behaviour (Peck and Tickell, 1992). While these levels have neither potential or meaning in isolation, they do allow a consideration of the potential 'intervention points' for action (Gibbs, 1996).

The existence of these levels of abstraction indicates that sustainable development will need to be promoted at a number of intervention points and at a number of spatial scales. Intervention at the level of regulatory forms may be easier to initiate and comprehend, but such concrete forms of intervention must be underpinned by complementary social values and norms i.e. the mode of social regulation as a whole (Flynn and Marsden, 1995). However, it is unlikely that sustainability can be promoted solely through intervention at one particular level. This is for two reasons. First, it is not possible to have a change at one level alone - change must come about *throughout* the mode of social regulation. Second, measures taken at any one level will be partial and have limited effectiveness and, indeed, may be counterproductive. Local strategies, for example, may well have a role, but these must be located within a supportive national and supranational framework. Each set of social practices that interact together to comprise regulation have their own set of key sites which vary in form and process.

Some are situated in economic space (e.g. the nature of the production process), some in political space (e.g. the extant pollution control regime) and some in social space (e.g. the local environmental milieu) (Goodwin and Painter, 1997).

In relation to the wider regulatory context it is possible that the current mix of international environmental agreements, growing public awareness of environmental issues, the rise of 'green consumerism', corporate environmentalism and the incorporation of sustainable development into local and national economic policy represent constituent elements of a new mode of social regulation. However, at present whether they will (or can) cohere into a mode of regulation is very much open to future shaping through social struggle and conflict. Much of the policy literature makes the assumption that this is already in progress, but this neglects the key role of the political processes involved. This depoliticization of ecological questions leads to the elimination of the constituent social components and consequences, such that environmental demands can seemingly be made, and implemented, in an unproblematic manner (Enzenberger, 1996). It may be more appropriate at present to see environmentalism as one of a number of alternative strategies of regulation or new collective wills which have the *potential* to have a radical impact upon the conditions of existence of a regime of accumulation. Thus the kinds of local environmental policy making regimes we have identified in this chapter will become either fully or only partially integrated into a broader regime of accumulation. However, we certainly do not see much evidence that environmental policy making regimes operating in Southern California are similar to those operating elsewhere, or that approaches like the NCCP will diffuse to a broader economic or political arena. Indeed, we question to what extent approaches like NCCP or UK local authority policy regimes represent truly integrative approaches to sustainable development, or simply sustain an environmentally destructive local mode of social regulation. That such modes may not necessarily be environmentally benign is indicated by the fact that certain elements of capital are already using the concept of sustainable development for the continuation of a particular set of social relations, as with the examples from England and Wales and Southern California presented here.

We may know little about how such local modes of social regulation are formed or operate. But we would suggest that the use of a reconstructed urban regime theory might assist in our understanding of local environmental policy making regimes. As Goodwin and Painter (1997, 22) indicate an urban regime as a site of regulation could be viewed as being situated at the intersection of political, economic and social space, involving as it does the combination of (local) state capacity with a variety of nongovernmental resources (some social, some economic). An investigation of these varying sites could well offer a fruitful avenue of research within the regulationist perspective.

In order to assist the development of practical and theoretical work in this field, there is much scope for further work.

Much theoretical reconstruction must be done. What is certain is that the basis of that theoretical reconstruction must be empirical research focusing on the concrete social practices of urban politics in specific places and times (Lauria, 1997, 8).

In a similar manner, there is an opportunity to investigate the processes at work in the development of local environmental policy making regimes. We have provided some tentative evidence to suggest that these exist and that they take different forms in different localities, but there is considerable scope for a more detailed analysis of such regimes.

Note

1   These regime types correspond with generic regime typologies developed by Clarence Stone (see Stone, 1989; 1993).

Acknowledgement

The research outlined in this chapter draws upon a number of research grants. The overall theoretical approach forms part of a project on Governance and Regulation in Local Environmental Policy Making funded by ESRC (Grant number R000237997). The UK empirical research outlined in the second section of the chapter was funded through the ESRC's Global Environmental Change programme (Grant number L320252132). Thanks are due to Jim Longhurst and Clare Braithwaite for their contributions to the research. The third section is based on research funded by the US National Science Foundation (Grant number SBR-9512033). This section draws upon research conducted with colleagues at the Department of Earth Sciences, University of California, Riverside. Thanks are due to Tom Feldman, Tom Scott and Jim Sullivan for their contribution to this research project. The authors gratefully acknowledge this financial support.

References

Angel, D., Jonas, A.E.G. and Theyel, G. (1995), Constructing consensus: environmental policy making in Massachusetts. *Urban Geography* 16: 397-413.

Cochrane, A. (1993), *Whatever Happened to Local Government?* Buckingham: Open University Press.

Cox, K.R. (1997), Governance, urban regime analysis, and the politics of local economic development. In *Reconstructing Urban Regime Theory: Regulating Urban Politics in a Global Economy* M. Lauria (ed), pp.99-121. Thousand Oaks, CA: Sage Publications.

Cox, K.R. and Mair, A.J. (1991), From localized social structures to localities as agents. *Environment and Planning A* 23: 197-213.

Elkin, S.L. (1987) *City and Regime in the American Republic* Chicago: University of Chicago Press.

Enzenberger, A. M. (1996), A critique of political ecology in T Benton (ed.) *The Greening of Marxism*, Guilford, New York, 17-49.

Flynn, A. and Marsden., T (1995), Guest editorial, *Environment and Planning A*, 27, 1180-1192.

Gibbs, D. C. (1996), Integrating sustainable development and economic restructuring: a role for regulation theory? *Geoforum*, 27(1), 1-10.

Gibbs, D. C. (1997), Urban sustainability and economic development in the United Kingdom: exploring the contradictions, *Cities*, 14(4), 203-208.

Gibbs, D. C., Longhurst, J. W. S. and Braithwaite, C. (1996), Moving Towards Sustainable Development? Integrating Economic Development and the Environment in Local Authorities, *Journal of Environmental Planning and Management*, 39(3). 317-332.

Gibbs, D. C., Longhurst, J. W. S. and Braithwaite, C. (1998), 'Struggling with sustainability': weak and strong interpretations of sustainable development within local authority policy, *Environment and Planning A*, 30, 1351-1365.

Goodwin, M. and Painter, J (1997), Concrete research, urban regimes and regulation theory in M. Lauria (ed.) *Reconstructing Urban Regime Theory*, Thousand Oaks, Ca.: Sage, 13-29.

Harvey, D.W. (1985),. *The Urbanization of Capital*. Baltimore: The Johns Hopkins University Press.

Hay, C. and Jessop, B. (1995), The governance of local economic development and the development of local economic governance: a strategic relational approach, Lancaster Regionalism Working Group Papers (Governance Series), 53 (Paper presented to American Political Science Association Annual Conference, August 30-September 1).

Healey, P. and Shaw, T. (1994), Changing meanings of 'environment' in the British planning system, *Transactions of the Institute of British Geographers*, 19(4), 425-438.

Jessop, B. (1997), A neo-Gramscian approach to the regulation of urban regimes: accumulation strategies, hegemonic projects and governance. In *Reconstructing Urban Regime Theory*, M. Lauria (ed.), pp.51-73. Thousand Oaks, Ca.: Sage.

Jonas, A. E. G. (1997), Regulating suburban politics: 'suburban-defense transition', institutional capacities, and territorial reorganization in Southern California. In *Reconstructing Urban Regime Theory*, M. Lauria (ed.), pp.206-229. Thousand Oaks, Ca.: Sage.

Lauria, M (1997), *Reconstructing Urban Regime Theory*, Thousand Oaks, Ca.: Sage

Logan and Molotch (1987), *Urban Fortunes: The Political Economy of Place.* Berkeley and Los Angeles: University of California Press.

Newby, L and Bell, D (1996), Leicester's lessons in local sustainability, *Town and County Planning*, 65(4), 101-102.

Painter, J (1997), Regulation, regime, and practice in urban politics. In *Reconstructing Urban Regime Theory*, M. Lauria (ed.), pp.122-143. Thousand Oaks, Ca.,: Sage.

Peck, J and Tickell, A (1992), Local modes of social regulation? Regulation theory, Thatcherism and uneven development, *Geoforum*, 23, 347-363.

Ramsay, M. (1996), *Community, Culture and Economic Development: The Social Roots of Local Action.* Albany NY: State University of New York Press.

Stone, C.N. (1989), *Regime Politics: Governing Atlanta, 1946-1988.* Lawrence: University of Kansas Press.

Stone, C.N. (1993), Urban regimes and the capacity to govern: a political economy approach. *Journal of Urban Affairs*, 15, 1-28.

11 Towards a Multisectoral Process for Environmental Protection and Sustainable Development in Rural Europe

HEINRICH SEUL AND CHRISTIANE WELLENSIEK

Introduction

Sustainable development is an enormous challenge for rural Europe. Rural areas in the European Union face two major problems. One is related to the cost of repairing environmental damage caused by the unsustainable use of natural resources. Cumulatively, across Europe this amounts to several billions of ECU per year. As these environmental costs of production are not reflected in entrepreneurial budgets there is little incentive to change production patterns (e.g. agriculture). The second issue is rural Europe, which covers 80% of the European Union, currently faces an exodus of its inhabitants. Poor infrastructure, high unemployment rates, few and unequal opportunities for jobs, severe educational constraints, lack of investments and little access to new information technologies encourage many - especially the young, to leave the country in search for better living and working conditions.

Today rural communities often have limited long term visions of how to develop and maintain their own distinctive cultural and social identity and encourage economic development. New development often represents a major threat to rural sustainability, and the delicate ecological balance of our cultural landscapes. But of even greater concern, is the increasing exodus from rural areas. This will have dramatic consequences for both the fragile ecological balance of our cultural landscapes and rural economy itself.

Developing sustainable, long-term economic and social perspectives for rural Europe therefore represents a major challenge as highlighted in *Agenda 2000*.

There is a need for a coherent strategy, which embodies a multisectoral process of integrating social, ecological and economic elements concerns of a variety of actors.

European Rural Agenda 21 describes a programmatic process which seeks to integrate the views of disoarative rural needs so that a strategy which provides local solutions, but also is sustainable can be developed. This paper presents some basic ideas for this multi-sectoral *European Rural Agenda 21*.

.

Rural Sustainability: Conceptual and Operational Problems

Within the context of rural development numerous sectoral solutions are applied to resolve numerous specific single issue conflicts. The need to maintain environmental qualities versus the need for resource consumption fuel; economic development; the income created through the consumption of natural resources versus the cost of environmental degradation. Often both of these perspectives are in conflict with local social and cultural interests. The needs of agriculture versus natural habitat or groundwater protection; land development for industrial, commercial or residential use versus agricultural or forestry use or by the decline of rural enterprises versus need for more rural jobs.

Such conflicts are often exacerbated by an acute separation of sectors, interests, goals, processes and influences. By working in an isolated and fragmented manner this sectoral approach is the main reason that the economic, ecological and social potential of rural areas is not fully realised. for rural sustainability to be achieved there is a need to ensure that these fragmented interests work together in partnership to develop new integrated perspectives linking profitable economic development, sustainable management of natural resources and, local participation in an local rural context. Visions on rural sustainability, policies for sustainable development in rural Europe, practical cross sectoral strategies to integrate the many conflicting rural interests, a solid and transferable methodology have to be developed.

In practical terms, a political and administrative framework is needed to contribute to rural sustainability. This provides an impulse and incentive to motivate and guide rural interest groups to start shaping the basic outlines of a common process in which rural actors such as local authorities, business and rural communities formulate their sustainability goals and are enabled to take active steps.

Achieving the Goal

In practical terms, a multisectoral, methodology programme is needed which helps to reduce private and public spending cover the costs of environmental destruction, enhances profits from sustainable development and initiates personal and local identification with the rural ecological, social, cultural heritage.

Personal perspectives must be generated, especially by the younger generations or rural communities which in turn can contribute to the goal of sustainability and at the same time generate new sources of income. Based on a crossectoral, bottom-up-approach, a *European Rural Agenda 21* could enhance multisectoral processes which enable rural actors to develop and define their own goals, strategies and activities towards sustainable rural development. The approach focuses on their very specific regional and local needs and conditions and helps to engender a sense of ownership from the local population, which is so important for implementation purposes.

Achieving the goal of rural sustainability is inevitably an enormously complex task. A great variety of actors, interests and issues, many of which are in conflict with each other need to be integrated to achieve a common goal. For sustainable development to be a realistic goal for much of rural Europe it is important that seven key elements are integrated into the process. (Figure 11.1)

A *European Rural Agenda 21* process brings together all important interest groups such as local and regional government bodies, private enterprises - particularly agriculturally related - producers associations, social, environmental and development organisations, rural women and others. The objective of the discussions is to develop a concensus about future options (economic, social, ecological) which are derived using environmental management and local sustainability. Each participating actor will be expected to develop their own sustainability goals and organise their efforts in a programmatic, measurable and fully operational way. Crossectoral co-operation should create positive synergies between the single sectors and thereby enhance competitive market advantage.

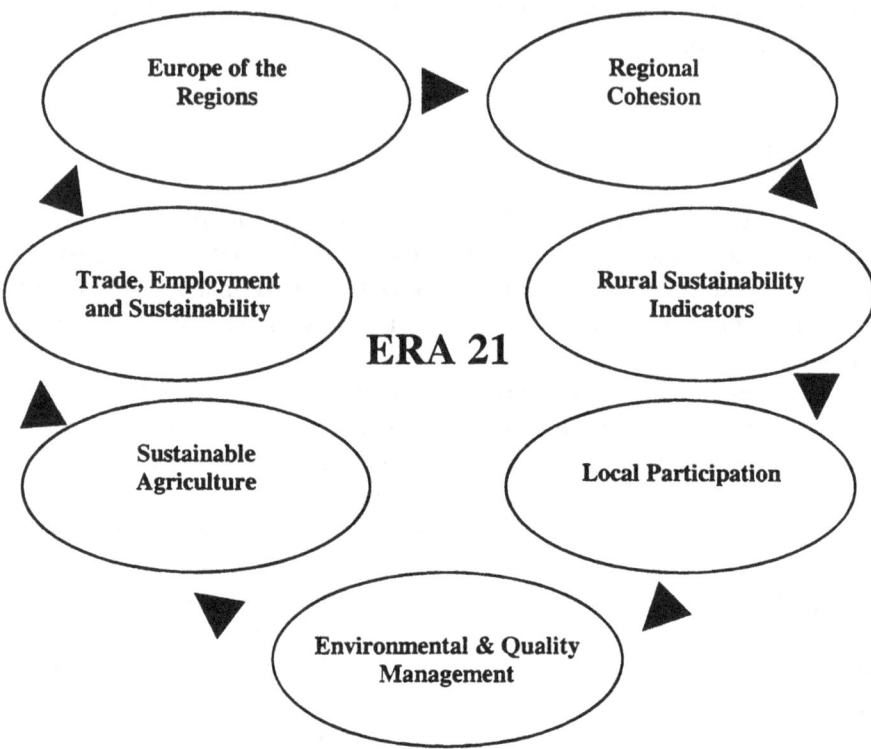

Figure 11.1 ERA 21 Key Elements of a European Rural Agenda 21

The seven key elements of a *European Rural Agenda 21* are outlined below

- *Europe of the Regions.* Regions will become more and more important in the process European integration and this will be reflected in its policies in relation to subsidiarity, cohesion and enlargement. In this context rural areas play a very important role within the future structural development of the European Union. This fact is being reflected by the growing importance of regional actors within the EU represented by specific bodies such as the Committee of the Regions;
- *Regional Cohesion.* Recent OECD (1997) research indicates that where regional centres experience positive economic development, their rural peripheries also tend to participate in economic success. It therefore follows by, fostering and focussing rural activities to enhance and organise co-operation between regional urban centres and their rural

peripheries is an important task for economic growth and stability. Collaborative work could involve helping to promote the area for inward investment and direct producer-consumer links could be (re)established. This latter project may help to identify and defuse local and regional environmental capacities which in turn may help translate the abstract concept of sustainability into something which is more tangible and can relate to individual actions;

- *Rural Sustainability Indicators.* There is a need to give greater weight to the real value of natural resources and the assets of rural areas. Techniques which could be applied include direct cost-benefit-analysis to patterns of resource use or environmental budgeting. New management principles such as sustainable flow management could be applied within the region. In making such adjustments to evaluation of options and management practices it will be necessary to identify key indicators which can be easily measured in a clear consistent and comparable manner in order to indicate the extent to which sustainability goals are being achieved;

- *Local Participation.* There is a need to initiate a bottom-up-process using the positive experiences of Local Agenda 21 processes. Only the full commitment and ownership of local actors and interest groups will enable locally derived sustainability concepts to be operationalised. It is likely that the key objectives for rural communities are going to be based around creating the climate for inward investment to enhance employment prospects, attracting new innovative investments and creating competitive market advantage for local producers;

- *Environmental & Quality Management.* There are new and growing range of tools designed to enhance local environmental quality management. Through a better utilization of resources and being seen to be more sustainable from the perspective of an organisation's management structure, not only helps to create a competitive advantage, but might also open up new markets. Applying formal environmental management systems (EMAS, the European Eco-audit Management and Audit Scheme or 1SO 14001, international environmental management standard), developing new and specific enviornmental agreements, or applying instruments to enhance environmental quality (promote agri-environmental goals or rural development), can all help to promote this agenda;

- *Agriculture Sustainable Key Industries.* With agriculture still a critical component of the rural economic, not just in terms of its economic and employment contribution but also in the way it manages the countryside from a landscape and ecological perspective, this industry can be arguably considered as the key industry for sustainable rural development. There are a range of programmes and initiatives that can be promoted. These include: developing agri-environmental services and markets, creating service-for-payment relations, enhancing organic farming / product marketing, input-output-balancing for the primary energy used, developing local and regional market concepts and applying agri-environmental indicators. Collectively such action can positively contribute to rural development.

- *Trade, Employment and Sustainability.* Notwithstanding the importance of agriculture, it is still critical to promote the diversificatiion of the rural economyby: stimulating innovation in environmental technologies, integrating product management and services, creating sustainable value added categories All these ideas, individually or collectively, can contribute to developing new marketing options, enhancing regional concepts, attracting new investment and venture capital and thereby creating new employment opportunities.

- *Realisng the potential.* By focusing on these key elements in a logical and systematic way a *European Rural Agenda 21* should help rural interest groups to promote their development needs in the context of sustainability by:

  - reacting to the growing importance of regions in the European integration process;
  - enhancing regional cohesion and co-operation between urban centres and rural peripheries;
  - developing specific, objective, applicable and measurable indicators for rural sustainability;
  - engaging all local actors in the process so that they have ownership, an involvement in decision making and a responsibility to take appropriate action;
  - applying environmental and quality management systems to as many organisations as possible;
  - developing concrete goals and an action plan which are appropriate for the needs of key local rural industries;
  - creating, and wherever possible, enhancing market synergies between trade, and employment with the concept sustainable development.

Conclusions

European Local Agenda 21 initiatives have tended to be dominated by urban development and environmental issues. These activities promoting environmental protection and sustainable development are very positive and to be welcomed. However, the need to include rural areas into sustainability processes is self evident. *Agenda* 2000 highlights the importance of this and recognises rural sustainable development as one of the major policy goals necessary to achieve successful and equitable European integration. In other words a strong rural initiative is needed to stimulate, govern and co-ordinate sustainability programmes and thereby advance socio and economic cohesion. The *The European Rural Agenda 21* concept can be seen as a process which can contribute to this task by enhancing economically attractive and socially authentic sustainability impulses from rural areas, which comply with the goals of Agenda 21. This concept should be further developed by all relevant interest groups in rural Europe and tested in practice within the framework of a European pilot project and/or the LEADER initiative.

Reference

Organisation for Economic Development and Co-operation (OECD), Paris 1997, *Sustainable Development - OECD Policy Approaches for the 21st Century*

PART IV:

EVALUATIONS OF PAST EXPERIENCE AND A VIEW FORWARD

12 Towards Coherence in European Spatial Planning?

DAVID SHAW AND VINCENT NADIN

Introduction

Throughout the 1990s, there has been a growing realisation that the general processes of European integration, and more specifically, many of the policies and programmes of the European Union, are having significant effects on spatial development. As a consequence such initiatives are inevitably impacting upon the spatial planning systems and policies of the fifteen member states (see for example, Davies et al 1994, Shaw et al, 1996; Williams, 1996; Wilkinson, Bishop and Tewdr-Jones, 1998). European Union policy initiatives such as regional policy, transport, environment, and agriculture all have important spatial objectives and impacts. In acknowledgement of this analysis, support has steadily grown for a more spatially co-ordinated approach towards EU policy and funding particularly at the supranational level, and more locally for greater collaboration between member states and regions in the preparation of plans and policies. Directorate General XVI (Regional Policy and Cohesion) has been encouraging this by focusing attention on both the scope for, and potential of, developing a European-wide and transnational spatial development framework. Two key documents Europe 2000 (CEC, 1991) and Europe 2000+ (CEC, 1994) have been instrumental in pursuing this agenda. The publication of the draft European Spatial Development Perspective (ESDP) in June 1997 provided the catalyst for further debate and action. The ESDP has been subjected to wide ranging consultation and debate and the final document was presented to the Informal meeting of Ministers of Spatial Planning at Potsdam in May 1999. The result is that European transnational spatial planning - the joint elaboration and co-ordination of spatial planning policy and action across national boundaries - has been evolving rapidly.

The increasing level of co-operative and collaborative activity across Europe is taking place in a context in which the EU has no competence in spatial planning, although its actions in the fields of regional policy, environment, transport and others do have major effects on national planning systems and policies. Furthermore there is enormous diversity between the member states in terms of the way their planning systems operate (CEC, 1997a). This diversity, deeply rooted in the historic, legal, cultural and constitutional influences of particular countries makes truly co-operative and collaborative activity difficult. The rationale for transnational strategic planning appears to be strong, well articulated and has widespread political support. However, the mechanisms, agencies and instruments that are needed to implement these ideas seem noticeably lacking. This chapter seeks to explore the growing raison d'être for transnational spatial planning and rehearses some of the arguments for promoting greater coherence between the operation of transnational, national, regional and local planning systems.

The chapter starts by reviewing the growing interest in transnational co-operation in spatial planning which appears to have gathered considerable momentum in the last few years. This is perhaps best exemplified by the emergence of the ESDP, and this is briefly described and considered. Moreover the influence of this policy document in shaping a national policy framework and the way it is beginning to be linked to European funding programmes, notably INTERREG IIC, raises some interesting areas for discussion relating to the competence of the EU to deal explicitly with spatial planning matters. What is emerging is a broad consensus of the need for greater vertical and horizontal collaboration in the field of spatial planning between both the EU and the member states and also amongst policy sectors. The precise mechanisms by which this may be achieved are still the subject of much discussion and debate.

An Explanation for Transnational Planning

One of the most important factors helping to explain the call for transnational collaboration on spatial planning is the growing economic interdependence of nations. Significant elements of economic activity together with political and cultural relations are effectively becoming globalised and independent of nation states. Whilst the depth and extent of globalisation can be debated, what cannot be disputed is that this process has specific and important implications for the changing patterns of spatial development. Within the European context some trends that have been noted include: an increased spatial concentration of economic activity combined with an enhanced central role for global and regional cities (Sassen, 1995); an intensified competition between cities (Parkinson et al, 1992; Harding et al, 1994; Newman and Thornley, 1996); the corresponding polarisation of economic prosperity between cities; the heightened differences between groups of citizens within cities leading too growing concerns about social exclusion; and, the negative environmental consequences of urban growth. Spatial planning and state regulation within the member states attempts to play a significant role in addressing these trends. Policies and plans attempt to maximise the competitive positions and growth potential of urban centres whilst attempting to ensure that, at best, the patterns of growth are sustainable and, at worst the negative impacts are avoided (Dieleman and Hamnett, 1994).

Whilst globalisation can provide a general context for the recent growth of interest in the transnational dimension of European spatial planning, four more specific explanations also need to be considered;

Spatial Impacts of European Policies

One of the fundamental goals of the EU is to achieve economic and social cohesion and balanced development across the Community. Such an overarching goal is evident in many EU policies and programmes and most pronounced in the sectoral policies related to agriculture, regional policy, transport environment etc. Such activities have very clear and direct impacts on patterns of spatial development. Many policy themes, including the promotion of social and economic cohesion or combating unemployment, have a very explicit spatial content by targeting resources at particular regions, areas or issues. This is reflected in many EU actions, three of which in particular are relevant for planning and can be used to highlight the relevant issues. The European Structural Funds attempt to counter the spatial disparities of unequal economic growth and economic restructuring across the EU. Much of the funding is targeted at specific areas or into the construction of infrastructure. In the most disadvantaged parts of Europe there is some evidence to suggest that such funding has helped the economic wellbeing of such regions by drawing them closer to the European average (CEC, 1996). The spatially discriminating nature of the Structural Funds provides a framework for national, regional and local planners and other groups involved in economic development and regeneration to use European money to shape planning policies and the influence the distribution of development activities. Second, the Trans-European Networks, (TENS) have been promoted with the specific aim of improving the relative location of peripheral areas within Europe. These networks are not merely concerned with transport, but also include access to energy sources and modern telecommunication systems. Many of the key European transport networks have been either focused in the more prosperous core areas or radiate to the periphery from the core. Thus improved access has been concentrated along corridors of growth and new pockets of relative peripherality quite close to the core may be emerging. Finally growing recognition of the need to reconcile environmental protection with economic growth particularly within the framework of sustainable development has provided spatial planning with new tools and responsibilities. Many of these have a strong European dimension. For example, the Birds and Habitats Directives have as their emphasis the protection of environmentally sensitive sites or the EIA Directives (CEC, 1985, 1997b) require the environmental impacts of large scale development projects to be taken into account in determining whether a project should proceed. Overall therefore, there has been an increasing recognition and interest in the spatial impacts and implications of EU policies programmes and projects (see for example Alden and Boland, 1996; Hardy et al, 1996; Williams, 1996). These might be considered direct impacts whereby the EU has shaped the process of planning by introducing new procedures, or imposed constraints as to where development might be accommodated or facilitated. In addition there has been a growing recognition that the actual process of European integration, has profound spatial ramifications. In trying to create a 'deeper' Europe, the Single European Market and more recently the prospects of Monetary Union, have and will continue to have, important implications for investment decisions and therefore spatial planning. At the same time, enlargement alters, and will continue to alter, the relative position of localities within European space.

European Spatial Planning Studies the European Commission

The first of these, *Europe 2000* (CEC, 1991) identified important spatial trends likely to affect the Community as a whole. This was perceived as an important step towards supranational planning providing planners with information and understanding of the wider processes and changes shaping spatial development and change at the European scale (Martin, 1992). In helping to promote the concept of planning in European space, *Europe 2000* divided the Community into eight transnational regions. This was based on the premise that these 'natural' regions exhibited similar characteristics and problems that could provide the basis for a more common and coherent approach. Whilst these mega-regions become the focus for further study, more importantly *Europe 2000* laid the initial foundations for transnational action on spatial development. In the forward to the document Bruce Milan, then the Commissioner for DG XVI said "it makes not sense for planning to stop artificially at national borders" (CEC, 1991, p3). Whilst *Europe 2000* was essentially a geographical text describing trends, the next key European document exploring the ideal of European wide spatial planning, *Europe 2000+* (CEC, 1994) went a little further. It began to evaluate the effect of a range of European Union policies on spatial development and also summarised the findings of the transnational studies. These studies, undertaken on behalf of the Commission by consultants, varied considerably in quality, character and the extent to which they moved beyond description to the promotion of positive action. Indeed in some of the transnational regions European funded programmes were established in an attempt in part to address some of the common issues affecting a region. For example the Atlantis Programme was specifically designed to promote transnational collaborative programmes in order to tackle some of the problems of the Atlantic Seaboard transnational region. The main role of *Europe 2000+* was to continue to promote the idea of the need for transnational planning and a broad European planning framework to co-ordinate action.

In order to achieve the wider Community goals of cohesion and competitiveness the central role of national planning systems and policies within each Member State was recognised (CEC, 1994). National and regional planning systems are and will continue to be the main vehicle for the implementation of broader transnational and cross border initiatives. While the planning systems themselves are a national responsibility and are likely to remain so, in order to promote co-operation there is a need for much greater comparable information about the way that the systems operate both in theory and practice. In 1993 DGXVI began the preparation of a Compendium of Spatial Planning Systems and Policies for the European Union, (Shaw et al, 1995). The Compendium describes for all the member states, the institutions and mechanisms which manage spatial development; describes planning policies pursued and through the use of case studies, attempt to understand how the formal procedures laid down in law and regulations function in reality.

In relation to the development of transnational planning initiatives, the Compendium projects highlighted a number of general points that need to be briefly identified here (CEC, 1997a);

- throughout the 1990s the planning systems in Europe have been in a state of flux. Most countries and regions are rethinking their approach to planning and making significant changes to systems or policies. There are some common trends and often the influence of Europe is important. For example, the requirement to have environmental assessment of all projects and the proposal to extend this to all plans and programmes affects all planning systems, although they can respond in different ways;

- there are changes aimed at improving the efficiency of the systems to regulate development, with, for example, concerted efforts being made to enforce planning regulations much more rigorously in the Mediterranean countries;
- there is a general strengthening of strategic planning, and more attention to mechanisms to achieve greater horizontal and vertical integration;
- there is increased recognition of the growing importance of the market and increasing reliance on private sector funding;
- in the regulation of development the apparently rigid link between legally binding plans and development decisions are being relaxed, particularly if the proposal is favoured by the determining body;
- there is also some decentralisation of responsibility for planning to lower tiers of government. This is particularly true of regionalised countries, and this may in turn lead to some fragmentation of approach for strategic matters within a particular country.

However the sheer diversity of approaches and differences of planning responsibility between different jurisdictional levels makes transnational collaboration difficult, though not impossible.

Political Support

This is a third factor giving rise to transnational planning with the emergence a 'Europe of the Regions'. The establishment of the Committee of the Regions (CoR), a consultative body comprising representatives from the local and regional authorities of the member states (Williams, 1995), provides an institutional arena where the regional dimensions of European policy can be openly debated. Indeed, the Commissioner for Regional Policy considers the CoR to be a vital consultative committee with regards to any European policy that has a very specific regional dimension (Wulf-Mathies, 1995). Such a view can be illustrated by the CoR's campaign during the debates that accompanied the revisions to the Treaty of the European Union (more commonly referred to as the Treaty of Maastricht). The CoR argued strongly for an increasing role and more activity in the field of European transnational spatial planning. In its submission to the debate the CoR stated:

> the Committee believes that the need to   promote cross-border and
> interregional co-operation between regional and local authorities should be
> the principle of spatial planning and spelt out in the Treaty in the interests of
> strengthening economic and social cohesion (CoR, 1995).

Perhaps the most important aspect of this statement is the call for an
explicit competence for the EU to deal with spatial planning matters. This
is an issue to which we will return later.

Individual Member State Actions

The actions of member states themselves have promoted transnational co-
operation. While the EU has no explicit competence to deal with spatial
planning matters, the member states have decided that this issue is
sufficiently significant to co-operate on an informal basis through
intergovernmental action. Any such action is based on an approach that is
consensual in nature and will be non-binding on the member states. At a
Ministerial level politicians in each of the Member States with special
responsibility for spatial planning meet regularly, at least once every six
months, to discuss issues relevant to spatial planning. Much of this
informal activity is developed through the co-operative action of the
relevant civil servants in the Committee for Spatial Development (CSD).
The CSD sets the agenda for the political meetings. This initiative includes
all the fifteen member states of the European Union and the European
Commission is also represented. A concern for a transnational perspective
in policy making has also been evident in the domestic planning
instruments and actions of some Member States. In 1989 the French
National Planning Agency (DATAR) published an analysis of European
spatial development which helped to promote investment in road and rail
infrastructure to reduce the perceived or potential peripherality of France
in a Europe whose centre of gravity was seen to be moving. The Dutch, in
their Fourth Report on Physical Planning, focused on maintaining and
enhancing the position of the Dutch economy in European space, and
similarly Denmark's *Towards the Year 2018* (Ministry of Environment,
1992) analysed Danish spatial structure within its European context. In
some areas of Europe there has been a longer history of co-operation
between member states. The Benelux countries, for example have a joint
vision or plan for their region, which has recently been revised. Also the
Baltic States whose joint work on *Vision and Strategies around the Baltic*

Sea (1994) identifies spatial planning goals and common actions needed to achieve them using spatial concepts such as 'strings', 'pearls', and 'patches'. In addition at the local level there are numerous examples of local co-operation which have led to cross border co-operation and collaboration often supported by the EU's INTERREG initiative.

Thus there are clearly a large number of growing and complementary actions and activities that help to support the notion of transnational planning activity. Critical points in the evolution of transnational planning have been the publication in June 1997 of the draft *ESDP* and the launch of the INTERREG IIC initiative. Both clearly promote the notion of the need to think and act strategically in spatial planning terms across national boundaries. They also raise a number of interesting questions relating to the need for the planning instruments between member states to be more coherent, and indeed, whether existing institutional arrangements are sufficient to ensure effective transnational collaboration.

The European Spatial Development Perspective (ESDP)

The first draft ESDP was published in June 1997 and marked an important step towards fulfilling the goal set by the informal meeting of Ministers responsible for spatial planning in 1993. At this meeting, held in Leipzig, there was political agreement that the member states should collaborate on a voluntary basis to try and progress beyond a series of planning studies towards the formulation of European spatial planning policy framework. Thus the ESDP is intended to provide this generalised spatial development plan for the EU, with the aim of providing European coherence and complementarity for the Member States own policies. Furthermore it was agreed that three basic objectives should underpin the development of the ESDP. These were: -

- the formation of a more balanced and polycentric urban system combined with the development of a new urban-rural relationship;
- parity of access to infrastructure and knowledge; and,
- prudent management and development of the natural and cultural heritage (CEC, 1997c).

The draft document was first made public in June 1997, following a political meeting of the CSD at Nordwijk, in the Netherlands. It has subsequently undergone further developments; a *'Final Draft'* was published in June 1998 following the European Council meeting in Cardiff; and extensive national and European consultation have helped to inform the a *'Final Approved'* document which was published at Potsdam under the German Presidency, in May 1999. One obvious criticism made about the development of the ESDP has been the long drawn out nature of the planning process. However as a document it represents a compromise between the different traditions, values and perceptions of spatial planning in the member states and required slow and often painstaking negotiation and discussion. For a fuller account of the evolution of the ESDP and the way that national spatial planning traditions have shaped the development of the document see Faludi and Zonnveld (1997).

Nevertheless, despite the various drafts and inevitable variations in style and structure, the underlying key themes and messages have remained remarkably consistent. It is a document that has been evolving gradually. It is not an easy document to read, though the need for greater transnational co-operation is clearly the most important key theme underpinning it. This is seen as necessary so that strategies and decisions at European, national, regional or local levels can be developed in a complementary way. This articulation of a policy objective, perhaps somewhat idealistically should lead to a greater synergy between transnational localities rather than unnecessary competition which is perceived as a wasteful use of resources. This complementarity needs to operate at all spatial planning levels and provides the key challenge to the effective implementation of the ESDP. There is also an implied recognition that instruments currently available at the level of the European Union, i.e. the legal instruments, (directives and regulation)and financial instruments are insufficient for the task in hand. The critical question then to be posed is can informal nature of the ESDP be sufficient for member states to modify their own instruments and policies thereby delivering the common trans-European policy objectives? The final section of the ESDP sets out a proposed framework for how the more thematic policy options identified earlier can be effectively integrated. This is perhaps the most significant part of the document. It represents a tacit agreement by all EU member states, and indeed the Commission itself, for a much more spatially integrated approach throughout Europe. Whilst the ESDP is not prescriptive about how this can be achieved, it acknowledges the need for:

- geographic integration at a European, transnational, regional and local levels;
- horizontal co-ordination between explicitly spatial and other sectoral policies; and,

- vertical co-ordination between administrative levels to ensure that policy options are complementary (CEC, 1997c).

Whilst the document is non binding on the member states, it has been made clear by the Commission and, *de facto,* by the effort that has already gone into producing the document that it must be more than a theoretical document. The ESDP will influence through persuasion, demonstration and incentives. Already there is close relationship between the principles and broad goals of the ESDP and the projects funded as part of the INTERREG IIC initiative (see below). Furthermore, it seems inevitable that the ESDP will become more influential. First, because the Ministers with a responsibility for spatial planning in each of the member states have agreed the content of the ESDP, they will wish to see their domestic policies conform to it. Already in the UK for example, the government is encouraging strategic planning particularly at the regional level to take much greater cognisance of the European dimension in developing strategy.   In 1998, the Department of the Environment, Transport and the Regions, (DETR) published a number consultation documents including *The Future of Regional Planning Guidance* (DETR, 1998). It was suggested that new regional planning guidance should use the ESDP as an important reference for strategic planning at the regional level. More recent guidance at the regional level recognises the importance of European wide policy and funding regimes in shaping and providing a context for regional spatial planning.

> In preparing Regional Planning Guidance therefore the RPB's and other stakeholders will need to take account of intergovernmental and Community based European activities policies, programmes and funding that impact on the region (DETR, 1999, para. 3.3).

Thus it is clear that in England regional planning in the future will have to acknowledge much more explicitly a European dimension to strategy formulation. Secondly it would seem inevitable that the current links between the ESDP and European Structural Fund programmes will be strengthened. Already the agreed operational programmes associated with the INTERREG IIC initiative are compatible with the strategic objectives outlined in the ESDP (see below). It also seems inevitable that the ESDP will influence the new round of structural fund spending due to be run from 2000-6. Because the ESDP represents a consensual view between the member states of how the European territory should be developed, it can be logically argued that key principles of this document can, and indeed have been integrated into the new Draft Guidelines for Structural Fund expenditure (INFOREGIO, 1999). Furthermore it would seem likely, and a logical extension to this, that in negotiating new operational programmes for spatially targeted European Structural Fund money, there will be an expectation that the strategies can demonstrate how they are consistent with and can contribute to the broad policy objectives outlined in the ESDP.

Whilst this is for the future, the early experiences of the way the INTERREG IIC programme has been operationalised in practice illustrates how an informal policy document such as the ESDP can be drawn more formally into formal European Union resource allocation processes.

INTERREG IIC

In October 1995 the Commission announced the third strand of its cross border co-operation programme. The previous and somewhat narrow geographical focus on border regions themselves was extended, through INTERREG IIC, to promote *transnational* co-operation initiatives in the field of spatial planning.

The INTERREG IIC programme is divide into three strands relating to transnational activities. The first part relates to transnational activities of a general nature and the second and third elements focus specifically on actions related managing the negative impacts of floods and drought. The overarching aim of the programme is to achieve a better balance better between the regions of the EU by encouraging an integrated approach to transnational problems. This in turn can be subdivided into three priorities:- fostering common transnational territorial development priorities; improving the impact of Community policies on spatial development; and, helping transnational regions take a more co-operative approach to the problems of floods and drought. Within this framework it is expected that there will be wide ranging discretion in the types of project funded, so long as they contribute to the general programme objectives.

Funding for the programme amounts to 412.84 million ECU. General transnational co-operation receives 120 million ECU and the remainder is allocated to the alleviation of problems associated with flooding and droughts (Table 12.1). Given the innovative nature of the programme the Commission entered into a round of information gathering and consultation with the member states and interested parties. An initial questionnaire survey and bilateral discussions formed the basis for a

Table 12.1   INTERREG 11c: The areas of cooperation and the funding allocated (1995 values)

	Transnational region/grouping	Countries involved	Funding
General transnational cooperation 121.03 mecu	Western Mediterranean & Latin (Italian/French) Alps	France, Greece, Italy Spain	13.30
	South West Europe	France, Portugal, Spain	5.15
	Atlantic Area	Ireland, France, Portugal, Spain, UK (3.75m)	13.03
	North West Metropolitan Area	Belgium, France, Germany, Ireland, Luxembourg (self-funding), Netherlands, UK (5m)	30.65
	North Sea Region	Denmark, Germany, Netherlands, Sweden, UK (3.2m), Norway (self-funding)	14.20
	Baltic Sea Region	Denmark, Finland, Germany, Sweden, Baltic States	24.28
	The Adrianube Region - the Adriatic, Danube, central and south-east Europe	Austria, Germany, Greece, Italy, the central and south-east European countries	20.42
Total for general transnational cooperation			**121.03**
Prevention of flooding 147.96 mecu	France - Italy		7.15
	Rhine - Meuse	Belgium, Denmark, France, Netherlands, Luxembourg, Switzerland	134.81
	Greece - Bulgaria		6.00
Prevention of droughts 144 mecu	Greece		16.80
	Italy		15.00
	Portugal		6.15
	Spain		106.05
Total			**412.99**

seminar attended by 250 people in July 1996. The purpose of the seminar was to formally launch the programme and facilitate working over the coming months. On behalf of the Commission, Mrs Wulf-Mathies in her address to the delegates stressed the innovative nature of the programme, which for the first time explicitly sought to organise 'transnational co-operation in the area of spatial planning based around concrete organisations'. The links to and consistency of the programmes to the ESDP was a common theme endorsed throughout the seminar. Finally three conditions for success were highlighted:

- the need for a small number of operational programmes so that the programme overall was manageable. Seven general transnational co-operation areas were established and these are complemented by a further six transnational projects funded by Article 10 of the ERDF;
- the need for each Operational Programme to identify a small number of themes; and,
- the need for a clear division of tasks between the Commission, member states and regional authorities who must work together in partnership respecting the principle of subsidiarity.

The UK falls within three largely contiguous transnational regions - the Atlantic Area; the North Sea Region and the North Western Metropolitan Area. These have parallels with the transnational study areas which first appeared in Europe 2000 and Europe 2000+ (CEC, 1991 and CEC, 1994) but with some significant modifications. The Atlantic Area covers the western parts of the UK and differs significantly from the Atlantic Seaboard by including much more extensive parts of inland and central France and Spain which had previously been associated with the Continental Diagonal. The North Sea Region has been more clearly defined, now comprising the eastern part of the UK and including parts of Sweden and Norway abutting onto the North Sea. The most significant change for the UK was the incorporation of the whole of the UK and the Republic of Ireland into a new region the North Western Metropolitan Area. This is a significant extension of the former Centre Capitals Region. This new mega-region was largely the result of lobbying from the Irish government who were unhappy at only being included in the Atlantic Area with its emphasis on peripherality. They argued that Dublin should more properly considered as part of Europe's metropolitan core. This extension of the former Centre Capitals Region means that large parts of the UK are eligible to participate in two different transnational INTERREG IIC Operational Programmes.

Following the definition of the transnational co-operation each partnership began to define its Operational Programme. These set out the broad strategic objectives of the programme, identify priorities, provide guidance regarding the types of projects that will be eligible for funding and describe the management arrangements for administering the programme. Once the Commission approved the operational programme, individual projects could be funded. Originally the operational programmes should have been submitted to the Commission in January 1997 but this was delayed. Eventually initial operational programmes for which the UK had a direct interest were submitted between March and July 1997. Following a period of revision and re-submission the operational programmes for the North Western Metropolitan Area were approved. There has been considerable delay in approving a programme for the Atlantic Area, but it did eventually get approval in June 1999.

Each project funded by INTERREG IIC must include partners from at least three member states. Partners may be regional and local governments, government agencies and NGOs. As with most EU programmes the need to attract co-financing is very important. The INTERREG IIC initiative has complex management arrangements the full implications of which have yet to be understood and evaluated. A Steering Committee with representation from both national and regional governments of the countries involved in a particular operational programme area will be responsible for allocating project funding. These bodies will be expected to come to decisions based on consensus. Programme and project evaluation will be the responsibility of a joint monitoring group made up of representatives from the countries within the co-operation area. Both groups will be served by a secretariat. For the North West Metropolitan Area this is based in the UK, for the North Sea it is based in Denmark, and for the Atlantic Area, the expectation was that it should be in France.

The ESDP and INTERREG IIC are ambitious experiments in transnational spatial planning and raise many interesting questions. There is a question about the level of local authorities interest in INTERREG IIC, not least because of the complex administrative and partnership arrangements and the relatively small sums of money available. Nevertheless, the dynamism of the planning systems noted earlier (CEC, 1997a), may be given further momentum from these transnational activities. As yet, while there is little demand for a harmonised system of planning throughout Europe, convergence trends are likely to continue, as co-operation increases. If borders are to become more permeable in land use planning terms then the planning institutions and instruments on either side of these political or administrative barriers must be able to communicate and act effectively together. In other words there needs to be greater coherence between systems. This does not necessarily imply the systems must all be the same, but is likely to involve all or some of the following:-

- the need for better understanding of the transnational aspects of spatial development issues;
- improved understanding of the capacity and constraints of systems and policies across state boundaries;
- amendments to existing instruments to enable them to embrace the transnational component of policy;
- modification to the existing institutional arrangements, or the creation of new bodies with a transnational mission; and
- consideration and resolution of the distribution of competences.

This final point raises further questions about where power and responsibility does and should lie for decision making in the field of spatial planning. Thus from a European perspective the issue of subsidiarity is a fundamental dimension of the debate.

Subsidiarity and the Transnational Planning Debate

Transnational action could be more easily facilitated if the member states established more formal institutions and instruments at the transnational level. It is important to make a distinction between co-operation and collaboration. The former has been a characteristic of networking activities and involves the sharing or ideas through lobbying working together for a common cause. It does not however require power or responsibility in decision making to be shared. Collaboration goes much further and implies a sharing of power. There are numerous examples in the field of cross border co-operation where the final step to collaboration has still to be taken. In the MHAL region for example, cross border co-operation between Dutch, Belgium and Germany local authorities, led to the creation of a loose strategy for the region. This has been recognised as a good example of cross border co-operation, yet still some of the more tricky and strategically important decisions, which are critical to a cross border region realising its full potential (e.g. the precise location of a regional airport) remain unresolved. Similarly the Kent-Nord-Pas de Calais co-operation is seen as being successful at one level. But even here several potential projects failed to materialise because of fundamental incongruities between the planning systems and the mismatch in the competences held by the co-operating partners. For example, much of Kent experiences considerable economic development pressure, which in turn generates considerable demand for good quality aggregates. There is a ready supply of aggregates across the Channel in the Nord-Pas de Calais region. It is therefore easy to construct a rational argument, applying the proximity principle, to promote the idea of cross border collaboration in this area. There are complications in progressing this idea because the jurisdictional incompatibilities between systems. In France responsibility for minerals planning lies with the national government compared to a delegated local competence within the UK.

If transnational planning is seen to be a good idea, and certainly recent evidence suggests that there is at least tacit support for this from the member states, then question of where power and responsibility should rest for the transnational element of decision making in spatial planning terms comes into focus. The subsidiarity principle may assist in this debate.

The Treaty of Union first enshrined subsidiarity into European law in 1992. There is much debate as to what it actually means. There are at least two basic interpretations of subsidiarity (Shaw, 1996 and Jones, 1997). The first is most commonly associated with truly federal systems where action takes place as close to the citizen as possible and a higher or central level authority only acts when it is appropriate to do so. For example the German constitution uses the concept of subsidiarity to allocate powers between the Federal and State governments. The second interpretation of subsidiarity is much narrower. In this sense subsidiarity is used to simply define the relationship between the EU and the member state. How the member state chooses to allocate competencies within its jurisdiction is really a matter for itself. In both cases the principle is not about decentralising decision making and autonomy down the government hierarchy, but rather a mechanism and principle which can be used to prevent power being ceded up to a higher authority unless it can be clearly demonstrated that there is a need for, and value can be added by taking decisions at higher jurisdictional levels.

The variation in the distribution of competences between jurisdictional levels amongst member states is likely to become a major factor in determining the extent and effectiveness of transnational collaboration in spatial planning. According to Armstrong (1997) there has been and continues to be a re-assignment of powers between multi-jurisdictional authorities. The term multi-jurisdictional is intended to be a neutral term encompassing both elected and non-elected bodies. He suggests that changes in assignment of powers can be altered by delegation of responsibility, the occupation of a function previously undertaken by another body or by altering jurisdictional boundaries. In many cases this involves both ceding powers up to higher tier authorities and perhaps devolving responsibility down the policy hierarchy. In this light the evolution of regional bodies and new public private partnerships offer new and emerging flexible mechanisms of governance to establish a strategic agenda for change. (Cappellin, 1997).

The complexity of the distribution of powers is certainly a challenge for further transnational collaboration. It is not merely appropriate for national, regional or local authorities to collaborate, it is important that they understand the power, responsibilities and autonomy of the prospective partners. The transnational dimension that is emerging is creating an ever increasingly complex multi-dimensional contest over power, resources and accountability. Much of this impetus is being shaped by intergovernmental co-operation. If such trends are to continue, then transnational planning may lead, through incremental change, to greater coherence in national, regional and local planing strategies between member states. This is not harmonisation per se, but rather steady convergence leading to greater coherence of a planning policies and actions. If transnational planning does not deliver the expected outcomes then there may be a case for extending the competence of the European Union to operate in the field of spatial planning and to ensure more coherence. This is a debate that is still to be had, but clearly EU impacts on spatial planning are only set to grow.

References

(CEC is the Commission of the European Communities or European Commission)

Alden, J. and Boland, P.(eds.) (1996), *Regional Development Strategies: A European Perspective*, Jessica Kingsley, London.
Armstrong, H.W. (1997), 'Regional-level jurisdictions and economic regeneration initiatives', in M. Danson, (Ed) *Regional Governance and Economic Development*, pp.26-46, London: Pion Ltd.
Cappellin, R (1997), 'Federalism and the network paradigm: guidelines for a new approach in national regional policy, Committee of the Regions', in M. Danson, (ED), *Regional Governance and Economic Development*, pp.47-67, London: Pion Ltd.
CEC (1985), Council Directive of 85/337/EEC of 27th June 1995 on the assessment of the effects of certain public and private projects on the environment. *Official Journal of the European Communities* C175:40-9, 5th July 1985.
CEC (1991), *Europe 2000: Outlook for the Development of the Community's Territory*, Luxembourg: OOPEC.

CEC (1994), *Europe 2000+: Co-operation for European Territorial Development*, Luxembourg: OOPEC.

CEC (1996), *First Report on Economic and Social Cohesion; Preliminary Edition*, Luxembourg: OOPEC.

CEC (1997a), *The Compendium of Spatial Planning System and Policies: Comparative Review.*, Luxembourg: OOPEC.

CEC (1997b), 'Council Directive 97/11/EC of 3rd March 1997 amending Directive 85/33/EEC on the assessment of the effects of certain public and private projects on the environment.' *Official Journal of the European Communities* L 073:5-15, 14th March 1997.

CEC (1997c), *European Spatial Development Perspective: First Official Draft*, Noordwijk, 9-10th June.

Committee of the Regions (1995), 'Opinion of the Committee of the Regions on the Revision of the Treaty on European Union', Brussels, 20 April.

Davies, H. W. E., Gosling, J. A. (1994), *The Impact of the European Community on Land Use Planning in the United Kingdom*, London: Royal Town Planning Institute.

Dieleman, F. And Hamnett,C (1994), 'Globalisation, regulation and the urban system,' *Urban Studies* 31(3) 357-364.

Faludi, A. and Zonnveld, W. (eds) (1997), 'Shaping Europe: The European Spatial Development Perspective,' *Built Environment*,Vol. 23, No 4.

Group of Focal Points (1994), *Visions and Strategies around the Baltic Sea 2010: towards a Framework for Spatial Development in the Baltic Sea Region*, Karlskrona: The Baltic Sea Institute.

Harding, A., Dawson, J. , Evans,R. And Parkinson, M. (eds) (1994), *European Cities: Towards 2000*, Manchester: Manchester University Press.

Hardy, S. Hart, M., Albrechts, L., and Katos, A. (eds)(1995), *An Enlarged Europe: Regions in Competition?* London: Jessica Kingsley.

INFOREGIO (1999), 'Brussels Forum on the ESDP' *INFOREGIO Magazine*, Feb 11th 1999.

Jones, J. (1997), 'The Committee of the Regions, Subsidiarity and Warning', *European Law Review*, 22 (August) 312-326.

Martin, D. (1992), 'Europe 2000; Community Actions and Intentions in Spatial Planning', *The Planner*, TCPSS Proceedings, 27th November 1992.

Ministry of the Environment (1992), *Denmark Towards the Year 2018: The Spatial Structuring of Denmark in the Future of Europe*, Copenhagen: Ministry of the Environment.

Nadin,V. and Shaw, D.P. (1998), 'Transnational Spatial planning in Europe: The role of INTERREG IIC in the UK,' *Regional Studies*, vol. 32(3) pp.281-99.

Newman, P. and Thornley, A. (1997), *Urban Planning in Europe: International Competition, national systems and planning projects*, London: Routledge.

Parkinson, M., Bianchini, F., Dawson, J., Evans, R. and Harding, A. (1992), *Urbanisation and the functions of cities in the European Community*, Liverpool Institute of Urban Affairs, John Moores University.

Sassen, M. (1995), 'Urban impacts in economic globalisation' in Brotchie, J. et al (eds) *Cities in Competition: Productive and Sustainable Cities for the 21st Century*, Melbourne: Longman.

Shaw, D., Nadin, V. and Westlake, T. (1995), 'The Compendium of European Spatial Systems', *European Planning Studies*, 3(3), 390-5.

Shaw, D., Nadin, V. and Westlake, T. (1996), 'Towards a Supranational Spatial Development Perspective: Experience in Europe', *Journal of Planning Education and Research*, 15(2), 109-116.

Shaw, J. (1996), *Law of the European Union* (2nd Edition), London: McMillan.

Wilkinson, D.. Bishop, K. and Tewdr-Jones, M. (1998), *The Impact of the EU on the UK Planning System*, Department of the Environment Transport and the Regions.

Williams, R. (1995), The European Union Committee of the Regions, its UK Membership and Spatial Planning. Working Paper 52, Department of Town and Country planning, University of Newcastle-upon-Tyne.

Williams, R. (1996), *European Union Spatial Policy and Planning*, London: Paul Chapman Publishing.

Wulf-Mathies, M. (1995), 'The future of EU regional policy' Speech given at the Centre for European Policy Studies, Brussels, 19 October 1995.

13 Non Grant Instruments and the Structural Funds: The New Alchemy?

PETER RAMSDEN[1]

Introduction

Non grant instruments or fund based approaches are the new alchemy of regional development. These are the approaches that promise to create more investment out of scarce public resources in the elusive inter-face between the public and private sectors. This paper attempts to identify why this has been a difficult path to tread, where the pitfalls are, and what lessons fund based approaches demonstrate for the wider use of non-grant instruments in the Structural Funds.

Non grant instruments described in this paper include a range of approaches to regional development based on revolving funds. They include, inter alia, venture capital, risk loans, interest rate subsidies and guarantee funds. The paper follows the Commission terminology in using the term risk capital to describe venture capital and loan approaches. More recently the revolving fund based approach has been applied in the fields of property and the social economy.

The first impetus for this approach came in the Delors White paper 'Growth Competitiveness and Employment' (CEC, 1994a) which emphasised a range of financial engineering[2] techniques, in particular the possibilities of venture capital as a motor for (small and medium-sized enterprise), SME growth and regional economic development. This stimulated a number of experiments in the 1994-96 Objective 1 and 2 programmes.

These early forays into risk capital have now been reinforced in the Commission's Agenda 2000 document (CEC, 1997a) which made an explicit call for non-grant instruments - including venture capital - to be developed within the context of the Structural Funds.

> The multiplier effect of structural resources should be increased by greater use of other forms of assistance than grants (interest-rate subsidies, guarantees, venture capital holdings, other holdings). (CEC, 1997a)

Due to a combination of circumstances, the 1994-96 programming period saw much of this new venture capital activity take place in the UK. Before this generation of programmes only Strathclyde had started experimenting with venture capital.   It was also a result of the paradigm shift that occurred in the UK's Structural Fund programmes in 1994. This was characterised by DGXVI at the time as a move away from the hard infrastructure of roads, bridges, and site development to the 'soft infrastructures' of business support, innovation and community economic development.

For the UK, these new approaches to venture capital offered an opportunity to step outside the straitjacket of public spending shortages epitomised by frequent complaints over the shortage of 'matching' funding or 'match'. Financial engineering offered a new way of involving the private sector in the programmes while avoiding the grant culture. There were also advantages in the programme delivery of this type of financial mechanism: Large numbers of enterprises could be reached through a single grant. This was ideal given the UK's relative lack of SME direct grant regimes and 'global grants'.

Risk Capital

The Commission defines venture capital in its Guide to Financial Engineering Techniques (CEC, 1994b) as a generic term covering the whole spectrum of investments in unquoted companies. In this paper risk capital is taken to include all types of investment that are not secured by collateral and includes both equity and loan instruments. A stricter definition of venture capital would restrict it to equity investment or 'shares'.   Under this definition, the types of debt financing known as risk loans in Finland or mezzanine finance in the UK, would be included under risk capital whilst not strictly being venture capital.

Investment finance can be divided between debt and equity. Debt financing is normally obtained from banks at fixed or variable rates of interest secured against the company's assets. The bank has the first call after the tax authorities, on any assets in the event that a company is liquidated. The returns on debt for the lender are never high and are usually set at a margin above the prevailing bank rate, arranged in such a way to offset the average cost of losses, plus administrative charges and profit. For medium-sized loans this interest rate is normally around 4% above base rates. Smaller borrowers pay higher rates or are unable to obtain credit because of higher risk. For large companies with correspondingly large loans the interest rates may be as little as 1% above base rates. The profits from debt financing come from volume and risk management. It is essentially a risk-averse business where the bank aims to offset risk by securing as much as possible of the loan through collateral in the company, normally in the form of property and equipment.

In the field of debt finance, the market gap that the projects funded through the Structural Funds have sought to fill has been in the area of smaller unsecured loans. These are the 'Risk loans', where no collateral is provided by the company and the financial intermediary uses public money to offset the risk. 'Mezzanine' finance is an extension of the risk loan concept where the finance sits between debt and equity by using a proportion of equity within the deal in the form of preference shares or 'equity kickers'.

Equity is at the opposite end of the spectrum from debt. Here, the investor takes a share in the company with the prospect of dividends and capital growth if the company does well and a zero return if the company does badly. It is usually the least secure form of investment and the last to be repaid in the event of company failure. Venture capitalists who take investments in unlisted companies expect a level of failure, but because of the high level of 'upside' contained in their deals, they have a good chance of making a profit out of the small proportion of their investments that do very well. By contrast, the banks have no such return on debt financing.

Companies use investment in different ways at different stages of their life cycles. Five clear stages can be identified:

- Seed Corn: Usually the finance for R&D at or before the establishment of the firm;
- Start Up: Normally required after a product has been developed, this is the launch investment needed to take the product to market;
- Expansion: Investment to help a profitable company expand and grow;
- Replacement: The investment used to change the balance between debt and equity in the firm, or to replace an existing investor;

- Management Buy-Outs: Investment generated when an existing management team - or in the case of buy-ins, an external team - raises capital to fund the purchase.

Public Assistance for Venture Capital

At the European level there is a perceived lacking of venture capital instruments for SMEs in comparison to the United States. This is particularly identified for technology based companies, and is accentuated in the less favoured regions that lack clusters of such companies to provide a critical mass of demand. This absence of a financial service infrastructure mirrors that of an advanced producer service provision in these regions. The Commission has attempted to explore this deficiency with the 1988 seed capital pilot project launched jointly by the Directorates for Regional Policy and innovation. Strathclyde was the only U.K. participant.

SMEs in the UK suffer two main problems in obtaining finance: high transaction costs and a lack of collateral. At particular times, such as the late 1980s, high and volatile interest rates have also been a difficulty. However, it is generally thought that if small businesses could access capital at the same sort of rates as large firms, they would be content.

For equity investment there is a lack of provision for investments below £500,000. The feasibility study for the Merseyside Special Investment Fund found that there were no providers in this range in the region (MSIF, 1994). Consultants in a similar study in Scotland found that the smallest investment made by Investors in Industry in Scotland during the previous year was £250,000 (Grant Thornton, 1996). Anecdotal evidence suggests that the main London based venture capital funds are reluctant to operate outside of the South East. The problems of the supply of capital are mirrored by a frequent weakness in demand because small firms are often resistant to giving up their equity to external purchasers.

The Structural Funds and Non-grant Instruments

Until the mid 1990s, the Commission had been cautious and ambiguous in its behaviour toward non-grant instruments. Their problem was that the Structural Fund regulations define the funds (ERDF, ESF and EAGGF) as grant instruments. Non-grant instruments - and loan funds especially -are therefore difficult to fit into the regulations.

The Delors White Paper in 1994 created policy momentum inside the Commission to do more with venture capital and try to resolve these problems. For the regional policy engineers in DGXVI, venture capital, was seen as more cost effective than SME grant schemes and business support measures while at the same time avoiding grant dependency. Revolving funds had the added advantage that they could contribute to exit strategies, and continue in to the future in a similar way that Marshall Aid had for Germany after the war. These factors, combined with the opportunities for private sector leverage, wide coverage and efficient delivery systems, made venture capital funds attractive.

Yet although there were many compelling arguments for doing more with risk capital, there were strong institutional forces within the Commission that had blocked change. DGXX, the Directorate General responsible for Financial Control takes the purest view with regard to expenditure of the Community budget. DGXVI, with responsibility for Regional Policy, is the custodian of the ERDF, tending to be more pragmatic and results-driven whilst remaining cautious over grey areas of eligibility.

Not surprisingly, the main pressure for liberalisation came from the geographic desk officers. Having led the negotiations with the regional authorities, the desk officers recognised the local demands to do something in this field. Desk officers were also pragmatic policy innovators, constantly looking to implement new types of actions.

Prior to this the rule was that only pure equity and interest rate subsidy schemes could be supported, because in these cases it was clear that the ERDF had actually been spent. In the case of equity the ERDF had co-financed the purchase of a share of the company. Interest rate subsidy was also eligible because - although administered as part of a loan package - it was objectively the same as a grant.

In contrast, loan funds could not be supported because the ERDF was effectively loaning to an SME through an intermediary and the grant was therefore never spent. There was also a second danger that EU expenditure would go into funds managed by intermediaries, and whilst appearing to appear to be spent for financial control purposes, it would actually be out of the control of the formal structures of the programmes. Indeed, under such a scenario the funds could even be used for ineligible purposes or spent after the programme had ended. For the strict interpreters of the regulations the argument was always that the ERDF was not a loan fund - after all, that was why the European Investment Bank had been created.

The result of these positions was that prior to the 1994-99 programmes ERDF rarely participated in the capital of venture funds. Instead, help was given to reducing transaction costs by using the ERDF to subsidise arrangement fees, third party guarantees and other revenue costs. Despite the regulations, several regions had used ERDF in Venture Capital funds before the SEM 2000 guidelines (CEC, 1997b). One example in the UK, was Strathclyde, a programme that at this point was a hot house of policy innovation. Their fund was public sector managed, and included ERDF in the fund's capital. Other examples included the PEDIP programme in the Portuguese CSF, and Finpiemonte in Northern Italy. Undoubtedly other arrangements existed elsewhere.

Resistance to the old orthodoxy began to break down with the introduction of the new programmes. During 1995 and 1996 DGXVI and DGXX had acrimonious and lengthy battles over the question of eligibility of expenditure as they tried to draw up guidelines for the SEM 2000 datasheets (CEC, 1997b). The most tortuous of these negotiations was over venture capital funds. At this stage, despite the different approaches taken within DGXVI over venture capital, a common negotiating position was taken that aimed at opening up the possibilities as far as possible. DGXVI pressed DGXX to change the guidelines so as to allow ERDF to form part of a risk capital fund.

The price for this change was a complex set of guidelines that ran to five pages. Included in these were special arrangements for approval that brought venture capital funds outside the ordinary programming procedures devolved to the Member States, instead requiring individual approval by the Commission services.

At first, the new guidelines were not made publicly available. This was a further recipe for confusion and delay. The guidelines were finally agreed with the Member States and published in May 1997, too late for the 1994-96 round of Objective 2 programmes.

The New Regulatory Framework

Revenue Models, Capital Models and the New Guidelines

The two ways that ERDF grants have been used with risk capital are called the revenue and capital models. Capital models place the ERDF in the fund itself, while revenue models use grants to subsidise transaction costs (and sometimes losses). The difference between the two can be defined in terms of the final beneficiary of ERDF support. In the revenue model it is effectively the individual SME, whereas in the capital model it is the fund itself. Essentially, the new guidelines opened up the way for capital models while making it less likely that revenue models would be supported in the future.

The New SEM 2000 Guidelines for Venture Capital Funds (VCFs)

The new Commission guidelines apply for capital models and covered the following main areas:

- Establishment: Articles of association to be provided, registration with regulatory bodies (normally IMRO in the UK) and clear separation between existing holdings and those co-financed by ERDF/

- Rates of Assistance: Maximum limits for investments in enterprises apply as the upper limits for VCFs. i.e. 50% of total eligible cost in Objective 1 and 6 regions and 30% in Objective 2 and 5b regions;
- Types of Investment: No investment in working capital. Investee firms must be SMEs. No investments in sensitive sectors (e.g. coal, steel, textiles);
- Involvement of Public Sector, Private Sector: Both the public and private sectors are expected to contribute financially. The guidelines require a private sector approach to fund management;
- The Procedure for Approval: All schemes outside the 'standard' model to go to the Commission for approval;
- Administration Charges: Administration may not exceed 5% of the sum invested in the fund. ERDF does not participate in administration charges;
- Community Policies: Funds must conform with all Community policies including competition policy. Any implicit grants (e.g. interest rate subsidy) over 'de minimis' levels must be notified;
- Closure of Fund: Clear arrangements must be made for the disposal of assets at the closure of the fund. These must be set out in the submission;
- Eligible Expenditure Period. The fund must be fully committed before the end of the payment period of the programme concerned (up to 2 years after the end of the programme). Any surplus will be paid back;
- Reporting. Funds should report annually to the Monitoring Committees.

The Merseyside Special Investment Fund

The application of the two models, involving revenue and capital, can be illustrated by the Merseyside Special Investment Fund, a new fund created on Merseyside during 1995-96 as a result of the Objective 1 programme. The fund started as a feasibility study in response to an outline in the Objective 1 programme document. It was included as a result of direct pressure from the private sector, both through orthodox channels in the form of the Chamber of Commerce, and through the megaphone diplomacy of a local pressure group, 'Manufacturing Challenge'.

Reflecting the tendencies of the time, the text of the programming document had specifically barred the use of ERDF in any loan funds. By the time that the feasibility study was being carried out, the new guidance note on risk capital was being discussed in the Commission, but the advice in the UK from Government departments, was that it would take too long to negotiate a change to the programme text. Instead, the consultants worked on a revenue based model for the two loan funds and a capital matching model for the equity fund. These three funds administered by three separate enterprises. In addition a fourth fund, the interest rate fund, which is not strictly a fund, but rebates interest on the mezzanine and the small firms fund is administered by the MSIF itself (Figure 13.1).

Figure 13.1 The Structure of the Merseyside Special Investment Fund

The structure of MSIF in three funds reflects the difficulties that its designers had in satisfying the regulatory challenge that the programme measure had created. The fund design was complex - particularly in the case of the two loan funds which were designed around a global loan from the EIB for the capital using the ERDF to mop-up transaction costs, losses and interest rate holidays. The small firms fund is described below to indicate the parameters (Table 13.1). Each fund was to have its own fund manager and a boxed-off set of company arrangements in order to protect MSIF Ltd, in the event of major losses.

Table 13.1 Components of the MSIF Small Firms Fund

Size of Fund £5 million
Size of loans: £3000 to £50,000
Fixed interest unsecured (about 6% over base)
1% arrangement fee plus capital holidays
25% failure rate assumed
Non revolving - principal returned to lenders at term
75% of losses covered by EIF
Guarantee for premium of 2-2.5% of loan
Interest rate rebate mechanism rebates 50% of interest annually if job creation and other targets are met.
Job target of creating 1500 permanent FTEs by end 2001 in 250 SMEs

The proposal was well received in the regional partnership and by the Government Office for Merseyside who sent it more or less unchanged to DGXVI for approval under the large project procedure. Protracted negotiations ensued with the Commission. The project was finally approved nearly a year later following lengthy consultations both between the services, with the project sponsors and government offices. Some key changes were made as a result including the addition of an interest rate rebate scheme for firms that achieved employment targets. The reporting arrangements to the Monitoring Committee were changed and new partnership arrangements were developed to involve the local authorities and other agencies on a more formal basis.

The funds themselves operate according to some basic principles:
- Money and Management: Use of non-executive directors in all Equity and some Mezzanine and SFF investments and these are integrated into Business Links;
- Public mission combined with private way of working and fund management;
- Political independence of fund managers;

- Investment Operational guidelines set by Investment Advisory Committee consisting of Barclays bank and partner organisations.

MSIF obtained the money to create the fund and finance the institution itself from a range of institutions. The investment funds total £25 million for direct investment, and were procured from the following sources:

- £5 million from ERDF;
- £5 million from Merseyside and Pilkington Superannuation Funds;
- £14 million from European Investment Bank (guaranteed by EIF) to be repaid in full after 10 years or on return;
- £1 million from SRB;

The two loan funds, the rebate mechanism and the central structure require an additional £11 million ERDF revenue support.

Table 13.2 Merseyside Special Investment Fund : Results to December 1998

Fund	SMEs Assisted	£ Million Invested	Leverage £ Million	Jobs Created	Jobs Safeguarded
Small Firms FF	201	2949	14948	891	465
Mezzanine	31	4680	17288	83	895
Equity	17	5.056	8.64	135	388
Total	249	12.68	40.87	1,109	1748
Target	500	25	45.0	2590	2140

Source: Government Office for Merseyside

In terms of results, the MSIF set out to achieve 2,500 'net' jobs over ten years in 500 assisted SMEs (Table 13.2). This would yield an ERDF cost per job of approximately £6,400, a figure considerably less than the average cost per job for the programme as a whole of £25,000. By late 1998, the equity fund had created 135 jobs, the mezzanine fund 83 jobs and the small firms fund 891 jobs. The small firms fund was achieving more than targets while the other two funds were both undershooting in the case of the mezzanine fund by more than 50%. The success of the Small Firms fund both in terms of deal flow and results led to an application for an extension to the fund, but by this time the new guidelines were firmly in operation and the model had to be adapted accordingly.

Following the Merseyside negotiation, three other Objective 2 areas included specific measures for Venture capital funds in their programmes. These were Yorkshire and Humberside; Greater Manchester, Lancashire & Cheshire; and East London. Other programmes - including the Highlands & Islands Objective 1 and the Western Scotland Objective 2 programmes - made provisions for venture capital funds within measures for business support. A summary table is shown below. (See Table 13.3).

Table 13.3 Funds where ERDF formed part of the Capital supported in Objective 2 programmes 1994-1996

Name of Fund	Characteristics	Fund Manager	ERDF Grant	Private Sector Contribution
Yorkshire Investment Fund	Single revolving fund; not for profit company	Yorkshire Enterprise	£2.5	£3.5
Airport Commercial Ventures Seed Fund	Small start up fund; equity only	Airport Commercial Ventures	£0.3m	£0.7m
North West Special Investment Fund	Equity only	Electra Invotec	£1.5m	£3.5m
BNFL Fund	Not for profit fund aimed at new technology based businesses	BNFL	na	na
North West Special Investment Fund	Support for existing funds	Enterprise	£1m	£2.5m
Western Scotland Investment fund	Support for existing fund	Public sector	na	na

Risk Capital: The Issues

The introduction of risk capital funds into the Objective 1 and 2 programmes has raised a number of issues, including: the time needed to obtain approval, difficulty with guidelines, the cost of money, grant rates, leverage and value for money. These are discussed below.

The Time Needed to Obtain Approval

The implementation of these funds was beset with difficulties. The length of time it took to negotiate agreement on the schemes was normally between six months and a year. For the project promoters it was a tortuous process, for the desk officers in regional offices and DGXVI a nightmare of complex guidelines, deadlines and uncertainty. The problems were by no means all at the Commission end.   Some projects were inadequately prepared and presented and issues to do with structure, benefits, rates of return had often not been adequately dealt with by the Government Regional Offices.

Later projects from the Objective 2 programmes had an easier time than the MSIF and were processed more quickly. This was because they adopted the more straight forward capital matching model and also because some of the issues had already been addressed in the first round.

Difficulty of Adapting to the Guidelines

Many project sponsors found it difficult to fit what they wanted to do into the guidelines. This particularly affected those schemes that attempted to combine 'money with management' by going beyond the conventional confines of fund management. Nearly all funds had to strip out business support activities and place them in separate free-standing projects to get beneath the 4% (later raised to 5%) administration criteria. This was possible where the support was relatively autonomous such as non-executive director schemes but much harder with more integrated approaches. It was also harder for the smaller funds dealing with tougher target markets such as the seed capital funds.

One result appears to be that project promoters have largely ignored the most risky end of the market - the start up and seed capital funds. The majority of funds appear to be targeted on established companies that are much safer investments. This may also be a function of the lower 30% grant rates allowed in Objective 2 areas.

A second problem arose with type of investment. The Commission guidelines do not permit investment in working capital. Yet increasingly, in a knowledge based society, plant and equipment are only a minority part of most company's investment needs. Developing successful innovation requires human capital and knowledge, thus incurring salary costs. The danger is that both the existing banking system and the new financial instruments are predicated on an industrial paradigm.

Grant Rates, Leverage and Value for Money

ERDF grant rates appear to be significantly better than with traditional grant schemes. This is partly because the maximum grant rate allowed under the Structural Fund regulations is lower for venture or risk capital schemes which are classed as investments in enterprises. In almost every case listed, the grant awarded was the maximum possible - 30% in Objective 2 and 50% in Objective 1. The loan fund elements of the MSIF are the only exception to the rule and in this case the grant awarded was even higher at 70% because ERDF does not form part of the fund itself.

Private sector leverage is higher than for other types of projects in the programmes. In all cases where the maximum was applied in Objective 2, the private sector contribution was seven ECUs for every three ECUs of ERDF. This gives a ratio of about 1:2. In the Objective 1 case of Merseyside, the ratio of public to private is 2.5:1, although this improves when 'indirect' leverage is taken into account.

The programme documents distinguish between private sector contribution and private sector leverage. Private sector contribution is the figure in the appropriate column of the programme financial table and in the declaration of expenditure. It also makes up part of the national co-financing of European grant (matching funding or 'match'). Leverage, on the other hand, includes not only the co-financing but also any other private funding that is a result of the investment. For instance, the MSIF insisted that it never supported more than 50% of a deal, the other 50% could come from banks, or other sources, would thus constitute leverage.

None of the schemes in the UK - with the sole exception of Strathclyde's - had significant public sector[3] co-financing funding. Several of the schemes claimed private sector leverage on top of contribution. Thus MSIF claimed a further £20 million of direct leverage and £25 million of indirect leverage (from other investors supporting the same firms) in addition to the contribution. Taking all of this into account leads to a gearing of 1:3 - an improvement on 2.5:1 quoted above. On the same basis the Objective 2 funds gaining rise to 1:4.

The Cost of Money

Within the risk loan funds, the question of the cost of money to the borrower has been a perennial debate between Commission officials and the project sponsors. Many of the projects have wished to avoid offering interest rate subsidy. Neil Kemsley, a board director of MSIF put this position bluntly in an article in the Financial Times:

> We are not here to give away cheap money. There is no point in backing businesses which are not commercially viable. They have to be sustainable in the long term (1997)

The Commission had a number of reasons for wishing to see low interest rates. Interest rate subsidy was an eligible way to spend ERDF and it was clear with a subsidy that the SME was the real beneficiary of grant aid and not the intermediary body. Interest rate subsidy could also be justified in the UK because base rates have been considerably higher and more variable than in continental EU Member States. Officials had also had recent and successful experiences of managing the European Coal and Steel Community soft loans where subsidies had been awarded in return for job growth.

The compromise achieved with the MSIF was to implement an interest rate rebate on the anniversary of the loan to firms that had achieved their financial and job targets. This had the effect of reducing the interest rate from around four to six percent over base rates to less than base rates. Only successful companies would benefit from the rebate. Three quarters of the hidden grant was paid for by Europe, the remainder coming from Single Regeneration Budget funds.

Other projects have had similar debates with the Commission arguing that cost of money is not the problem. What SMEs need is the same access to money that larger firms can obtain at the same price. That said, large companies can often obtain debt financing at only one percent over base rates – a much lower rate than that proposed for most of the loan schemes applying for ERDF funds.

Intermediaries

Nearly all of the schemes involved the use of professional Fund Managers to select investments, provide support and administer the funds. There is a risk with these types of funds that the main beneficiaries will be the financial intermediaries and consultants involved in the schemes rather than the SMEs themselves. The Commission sought to control these costs by insisting on a ceiling of 5% for management fees

Taking again the MSIF as an example in its revenue-funded models, the price of arrangement fees (fees paid by the SME to the intermediary for third party legal and accounting expenses) were in the region of 10% of investment in the business plan. ERDF is then used to bring this down to an affordable 4%. These figures are considerably in excess of the industry norms of 1-4% depending on size of investment. Similar ratios were found for management fees. There appears to be an in-built tendency within revenue models for project promoters to enhance the fee income and use the ERDF to reduce it to bearable levels for the SME. For this reason alone, the capital model appears to be preferable.

A second area of concern is the degree of profit taking for fund managers and private investors. In the capital models, the fund managers earn a share of the final receipts in addition to an annual fee. In the MSIF Equity fund the fund manager would make a performance related fee worth about one fifth of the funds in management if the investments performed at the level anticipated by the business plan. The debate here is whether it is acceptable for private firms to achieve such high returns on work for public programmes. The flip side to this debate, of course, is that many public projects receive considerable grant aid for no results at all.

The investors themselves will also stand to gain good returns and have the risks of losses protected by the ERDF. In the MSIF equity fund model the continuation fund will receive no funding until the private sector investor has received back all of its original investment, plus a ten per cent per annum return. This means that, in effect, £5 million of public ERDF has to be written off before the private sector loses money. In the event of profit gains it is only over a 20% rate of return that profits are equally shared between the parties. For the private sector, therefore, it is a one-way bet.

The final area of concern is over set-up costs. Most of the funds did not receive money for feasibility or project design phases. MSIF was the exception to the rule where an initial grant of £100,000 of ERDF was made for the design phase and is rumoured to have over run the budget considerably. Other consultancies on Merseyside have grumbled that the set-up contract was never tendered and that the rewards were excessive.

There must be concern about the level of fees and profits built into some of these schemes. If too much money goes to the so called 'fat intermediaries' less can go to the beneficiary SMEs to cover risk, or reduce interest rates. There appear to be two possible safeguards. Firstly, there could be a separation of powers between the fund and the fund manager. This has been the case in some of the UK schemes – notably MSIF - but not in others. Secondly, an insistence on public tendering for the carrying-out of the fund management function would expose the fund managers to open competition, thereby reducing the charges for management. The case where a private venture capital company applies directly for Structural Funds assistance appears to be fraught with dangers and is unlikely to have sufficient checks and balances to safeguard the public interest.

Regional Institutional Capacity

What impact did the new funds have on the regional institutional capacity? Although it is too early to answer this question, there are some clear indicators. Only two funds - both managed by Invatec - brought a genuinely new player to the regions in which they were based. It is certain that Invatec would not have made a start in Merseyside without this backing, although the position in the rest of the North West is unclear.

The other funds served to strengthen the role of the existing bodies within their regional catchments. In most regions, DGXVI insisted on single funds for the whole region (Yorkshire, West Scotland, Merseyside). However, this rule broke down in the North West region (excluding Merseyside) where four funds were permitted with only a weak co-ordinating mechanism to link them.

Only MSIF set out to establish a totally new regional financial intermediary with co-location for its three fund managers in a single office. The Commission press notice glowingly described it as "a new regional financial institution, created on Merseyside for Merseyside firms".

Conclusion – Towards a New Fund Based Approach

Venture Capital and Risk Loans - Lessons for Policy Implementation

If we take alchemy to mean a 'the means of achieving a miraculous transformation'[4] then non grant instruments have certain characteristics that are relevant not least in their requirement for miracle workers in the form of project promoters, consultants, and officials.

Venture capital and risk loan funds were the first of the non-grant instruments to be tested in Structural Fund programmes. The experience has been mixed and has involved a long and sometimes painful learning curve for these 'miracle workers'. Public agencies will have to take more risks of this type in order to develop the flexible funding models of the future. However, this risk-taking will need to be backed up by more research and evaluation and dissemination to enable a better understanding of the issues - such as financial displacement, effects on recipient SMEs, employment opportunities etc.

There are several other lessons for those involved in Structural fund programmes. Among the programme secretariats there was clearly insufficient understanding of these types of projects. Many of the projects had been approved by regional secretariats without sufficient understanding of the ERDF guidelines and thus in effect were not ready for approval. Non-grant instruments are more complex than ordinary projects. Important issues of eligibility, rates of return, fee levels and institutional structure need to be addressed. There is, therefore, a growing need for training and exchange of experience to help increase understanding and raise the standards of implementation.

The Commission also found the projects difficult to process. Without more speed, transparency, co-ordination and dissemination, future moves to innovate in this way will be fraught with difficulties. Implementation was hampered by uncertainty over rule systems, the delays in publishing the SEM 2000 guidelines and the difficulties over their interpretation. Risk capital was also an example of where the expertise of the Commission was never co-ordinated. The implementation of new policy would have been assisted through a more collegial task force approach that could bring together internal expertise - DGIII (Industry), DGXXIII (SMEs), & DGXIII (Innovation) – with external expertise, for example from the European Venture Capital Association, to give the desk officers and secretariats support. Instead, divisions and departmentalism led to a failure to capitalise on skills and experiences both in the Commission and outside.

There has also been a lack of dissemination about best practice experience within the field, of evaluation of the early projects and of capitalisation on the experience that has been gained.

Risk Capital Projects: Alchemy in Policy Results

Despite the many problems in implementation, there is evidence that the new structures being developed at regional level for risk capital are making a valuable contribution to economic development and job creation.

MSIF was one of the more successful projects with publicly available data. One surprising finding emerging from the results available suggest that risk loans – and especially those at the smaller end of the scale, may be more potent at creating jobs than equity investment or pure venture capital. This finding would be against the conventional orthodoxy which argues that it is equity where the main market gap exists.

It may be that ensuring adequate supply of debt or loan capital is a more appropriate use of Structural Fund support than venture capital. Risk loans fits better into a programming framework than venture capital which by definition is something of a lottery. In addition there is a case to be made that risk loans are a more manageable product than venture capital, because they is simpler. Venture capital support requires decisions on questions of risk and return that seem to be very difficult for public bureaucracies to take.

New Types of Non Grant Instruments

For the future, a number of new sectors and players need to be considered for non-grant instruments.

- The greater use of revolving equity and debt instruments funded under ERDF for infrastructure projects through property funds and for supporting the third sector through social capital funds.
- New types of public/private financial intermediary to account for the funds and control fund managers that combine public mission with private ways of working.
- New ways of working with the European Investment Bank and European Investment Fund for all types of fund based approaches, bringing the economies of scale and security of these institutions to the regional level through more flexible global loan and guarantee arrangements.

New Types of Intermediary Bodies: Front Office – Back Office

These new financial bodies are carrying out an intermediary function that separates the complexity of funding, monitoring and reporting into a 'back office'. All the difficulties of bringing together complex funding regimes and coping with bureaucracy stop here. The 'front office' is where the consumer – in this case the applicant SME - is dealt with in a quick and businesslike way without any contact with funding bodies ERDF application forms or lengthy delays for the decision or payment. Fund based models offer a powerful mechanism to bring Structural Funds closer to the consumer without strangling them.

There is considerable scope to replicate front office/back office approaches in a 'fund based approach' in other parts of programme delivery where there are a large number of recipients. Community economic development priorities would be an obvious target through non-grant instruments to finance micro credit and local social capital approaches. New intermediaries would need to combine a public sector ethos about value for money, objectives and evaluation with a social economy approach to managing and working with the communities that they are aiming to help.

Fund based approaches do offer a real opportunity to engage the private sector and to design projects that can contribute to regional development. They are highly complex and present major challenges to the bureaucracies that approve funding. Like all other alchemies they are not magic. Like other policies they have to be well designed and crafted to produce good results.

Notes

1   The author worked as a detached national expert for the European Commission from 1993 to 1996

2   Financial engineering is described by by the Commission as "at its most basic the process of developing solutions to financial problems " and more generally as a shorthand for techniques involving risk or venture capital. It has other meanings outside European Funding and is often used in the City of London in a derogatory way to describe creative accounting.

3   Public sector expenditure is defined here as spending that is entered in the national accounts

4   Oxford English dictionary defines alchemy as seeking to turn base metals into gold or silver; miraculous transformation or the means of achieving this.

References

CEC (1994b), *Guide To Financial Engineering Techniques Used By The Commission In The Context Of Regional Policy.*

CEC (1994a), *Growth Competitiveness And Employment: The Challenges And Ways Forward Into The 21st Century.* [Delors] White Paper, Brussels : CEC.

CEC (1997a), *Agenda 2000: For a stronger and wider Union CEC,* Brussels : CEC.

CEC (1997b *EU Structural Funds: Eligibility expenditure Datasheets.* Brussels.

Fazey, I.H. (1997), Merseyside: Smaller businesses find a saviour fund, *Financial Times* 18th December.

Grant Thornton Scotland (1996), *Loan and Equity funding for SMEs in Central Scotland.* Final report to Eastern Scotland European Partnership and Strathclyde European Partnership.

MSIF (1995), *Merseyside Special Investment Fund Business Plan* Ernst and Young.

14 European Union Regional Programmes - Lessons from Practice and a Review of Future Options

PETER ROBERTS

Introduction

The Regulations governing the operation of the European Union's Structural Funds (Commission of the European Communities, 1996) were drafted with the intention of providing a common basis for the operation and management of regional programmes throughout the fifteen member states. In theory, this common approach should provide a uniform basis for the implementation of the Structural Funds and should facilitate a comparative evaluation of the performance of the programmes. However, in reality, within the context of EU-wide guidance and regulations, fifteen or more approaches and systems of operation have emerged. Even within a single member state it is not unusual for a range of systems to exist. In such circumstances the participants involved in the processes of developing and managing programmes may be required to engage in various styles of partnership behaviour and may be required to perform a variety of different roles depending upon which region they are operating in.

Variations in the operational and management styles of regional programmes are to be expected in any policy system that attempts to mesh together the regional planning and development structures and aspirations of fifteen member states. In one sense, the fact that a common policy system, together with an associated method for its implementation, still continues to function at all is a tribute to the collective will of the various partners involved in the operation of the regional programmes. However, in many cases this collective will has to be translated into operational management and development "on the ground" through complex, and often fraught and tortuous, systems of negotiation and compromise. Whilst the end result may be judged by outsiders to be worth the effort involved, the question should be asked: could a similar result be achieved with greater economy in terms of the resources devoted to development and management, and to ensuring the compliance of programmes with the Regulations?

This question, together with a range of other areas of investigation, formed the basis for research conducted initially in the UK and, later, in Belgium, Germany, Ireland and the Netherlands. The research undertaken in the UK was funded by the Joseph Rowntree Foundation, whilst the work outwith the UK was supported by a research grant from the Central Regional Council.

This chapter provides a summary of the findings from this programme of research. In the following section of the chapter a context is provided for the more detailed evaluation which is presented in section three. Section four offers some initial conclusions regarding the design, operation and effectiveness of the regional programmes, and it also offers some thoughts on the future construction and implementation of the programmes supported by the Structural Funds.

As noted above, this chapter reports the results of two linked projects that form part of a long-term programme of research into the origins, development and operational characteristics of the programmes supported by the Structural Funds, and the likely future evolution of European spatial planning and development policies. This chapter draws on five working documents that were prepared during the early stages of the research programme.

The Context for The Current Regional Programmes

Background

Even though the Treaty of Rome made reference to the existence and significance of regional disparities and problems of restructuring, it did not contain specific powers that allowed either the (then) European Economic Community, or individual member states to take action to address such problems on behalf of the Community (Armstrong, 1989). However, during the late 1960s research undertaken by, and on behalf of, the European Commission identified the case for intervention. This growing recognition of the need for a form of Community-wide selective regional assistance, together with the desirability of introducing a more explicit spatial dimension into a range of other policy areas, resulted in the establishment of the European Regional Development Fund (ERDF) at the Council meeting held in December 1974 (Roberts et al, 1993).

The ERDF came into operation in 1975 and, within the context of a set of principles governing the operation of the Fund, provided additional financial support for regional policies and programmes that were (in most cases) already well-established. Most of the pre-existing national schemes of regional assistance provided financial and other support to public authorities and firms in designated areas. In the early years of the ERDF financial support was allocated through a system of national quotas. Under this system the designation of areas and the primary assessment of applications for assistance were undertaken by the responsible authority within the member state (usually either central or regional government). Key weaknesses of the ERDF under the quota system included: the limited degree of strategic guidance that could be exercised by the Commission over the use of the financial support made available, considerable variations between member states in the definition of assisted areas and the use of indicators for designation, various interpretations of the additionality requirement, the lack in many cases of consistency in the use of financial support between sectors, and the absence in most cases of any rolling programme for regional development. The result was a policy system (or systems) that has been described by Armstrong as an "extraordinary patchwork" with member states dictating the implementation of supranational policy due to the "strangle-hold" which they exerted "over who got ERDF assistance" (Armstrong, 1989, p.172).

Whilst the quota system did, in theory if not in practice, allow for a flexible mode of implementation of Community regional policy which was tailored to the requirements of individual regions, it "prevented the EC from concentrating all the effort on the poorest regions" (Vickerman, 1992, p.49). The reforms agreed in 1984, and which were implemented in 1985, introduced the principle of programme funding and replaced the system of national quotas with a new financial distribution mechanism based on indicative ranges. Many of the innovations introduced in 1984 were developed from the experiments in programme funding that had been encouraged under the non-quota element of the ERDF, established in 1979. During the period from 1979 to 1984 an experimental system of non-quota funding operated. Some 5 per cent of the ERDF total resource was allocated in support of designated programme contracts that were multi-annual and, chiefly, targeted upon a specific set of measures designed to address the problems associated with the decline of a local dominant industry. The proven value of this experiment in programme funding provided substantial evidence needed to justify a move away from the quota system.

The post-1984 system merged the quota and non-quota elements of the ERDF (although over the short-term many of the distinctions between the two elements persisted) and, more importantly, it tied assistance to an approved regional development plan. In addition to the main areas of funding activity, the post-1984 ERDF also introduced Integrated Development Operations (IDOs) as a significant feature of the policy portfolio (between 1979 and 1984 two experimental IDOs in Naples and Belfast, had been designated). The IDO model was designed in order to help redevelop and regenerate run-down urban areas, and was based on partnerships in funding and management between the EC and national, regional and local governments.

A further reform of the policy regime for the support of regional restructuring took place in 1988. The 1988 reform of what became known as the Structural Funds brought together the ERDF, the European Social Fund (ESF) and the guidance element of the European Agricultural Guidance and Guarantee Fund (EAGGF) with the aim of concentrating resources on the achievement of five objectives and in accordance with six principles. The first full round of programmes operating under the 1988 Regulations commenced in 1989. This reform firmly established the programme principle, which had been introduced in 1979, and it set out specified working methods for the design, development and implementation of regional programmes.

Under the new procedures, member states are required to submit regional development plans and/or strategies prepared and agreed by partnerships representing a range of national regional and local interests. At this point it is important to note that although the Commission is a member of each of the partnerships responsible for managing approved programmes, in reality the Commission performs two roles: as the custodian of Community interests and as a partner in individual programmes. Following their preparation at national/regional level the regional plans/strategies are then the subject of a process of negotiation which results in the formulation of a Community Support Framework (CSF). This represents an agreement between the Commission and an individual member state, and it outlines the purposes and targets to which both European and other financial resources will be directed. Within CSFs more specific actions are defined in the form of Operational Programmes (OPs) and other measures including the provision of global grants to designated authorities. In the Cohesion countries both the Structural and Cohesion Funds are planned together within a single CSF (Bradley, 1995).

The procedures, principles and practices that were expressed in the 1988 Regulations were reviewed during the early 1990s and a package of revisions was agreed in 1993. Compared with the changes made in 1988, the 1993 reform was minimal. A new Objective 4 was created in order to assist in the adaptation of workers to changes in industrial structures and systems, and a variety of procedural charges were introduced in order to simplify decision making. The accession of Sweden, Austria and Finland to the EU marked the addition of a new Objective 6 which has the aim of assisting in the adjustment of regions with extremely low population density.

This brief introduction to the emergence and evolution of the regional programmes illustrates the varied origins and sources of agreement which have helped to frame both the broad rules and the more formal principles (some of which are incorporated in the formal Regulations, whilst others are not) which govern the current operation of the Structural (and Cohesion) Funds. This distinction between formal Treaty obligations and Regulations, on the one hand, and the broader rules of engagement, on the other, is of particular importance in an analysis of the practice of regional planning, development and management. Points of particular interest and relevance in the context of this chapter include:

- the longstanding nature of many of the basic principles that guide the overall operation of the Structural Funds, even though some of the principles are observed more in name than in substance;
- the emergence of the current system of control and management as a result of experimentation and the limited evaluation of previous rounds of policy - this has led to additions or deletions on an incremental basis, rather than the fundamental restructuring of the central core elements of the system;
- the continuing strength of the programme approach - this is considered to provide an essential spine for the design and implementation of measures designed to address local and regional disadvantage, and it has the particular merit of allowing the identification of actions and measures that are mutually reinforcing and which involve the co-ordinated input of resources from a number of partners and other organisations;
- the progressive and constant addition of new operational roles and tasks to the programmes supported by the Structural Funds, in some cases a regional programme may be an unsuitable or unsatisfactory vehicle for

the discharge of such tasks, but, in the absence of other options, the task is, nevertheless, added to an already overburdened programme - for example, the regional programmes are now expected to deliver environmental sustainability, political cohesion etc;
- the increasing importance of the interface between the traditional role of the regional programmes and new or enhanced policy areas such as the recently strengthened environmental policy of the EU, or the rapidly emerging concern with matters of spatial planning and management.

This brief review of the historical and institutional background to the current regional programmes illustrates the importance attached to the operation of a policy system that is designed to reduce disparities between member states and between regions within the member states. For further background material on this aspect of the chapter see Albrechts et al (1989); Armstrong (1989) and (1994); Michie and Fitzgerald (1997); Hardy et al (1995); Roberts et al (1993) and Vickerman (1992).

Theoretical and Institutional Context

Although there is little that can be offered in terms of theoretical justification for the particular way in which the original ERDF was designed and structured, the establishment of a capacity to deal with the occurrence of regional problems is based on a much more substantial body of theory. The economic arguments in favour of a common regional policy are well known, Armstrong (1989) summarises them as:

- European regional policy can help to improve the overall efficiency of regional policy by ensuring that spending is concentrated where it is most needed;
- there is a need for greater co-ordination between national and regional policies in order to prevent economic and organisational chaos, reduce the possibility of competitive outbidding, and assist in the design and implementation of a common strategy for cross-border regions;
- a common interests argument can be advanced which reflects the consequences for member states of allowing regional problems to persist in fellow member states;
- a final argument is that regional disparities are a significant barrier to further political and economic integration.

These economic arguments are extended and reinforced by other organisational, political and spatial concepts. Three broad areas of theory provide the underpinning for a framework of analysis that allows for the development of a better understanding of the operation of regional programmes: organisational theory related to the structure and operation of modes and methods of policy formulation and regulation; political theories underpinning the extension of integration within the EU and related to the definition and exercise of subsidiarity; and spatial theories related to the designation and functioning of regions as appropriate spatial units for the co-ordination and integration of relevant aspects of public policy and for the design and implementation of development programmes. Each of these areas of theory is discussed below.

Organisational theory related to the design and operation of policy systems suggests that the tensions which exist within an organisation can provide an incentive to improve overall performance and the quality of the range of services provided. However, if left unchecked, these same sources of competition and tension (for example, core versus periphery, insiders versus outsiders, etc.) can lead to the fragmentation of implementation and can cause a decline in performance. A clear parallel exists between this broad organisational concept and the behaviour of institutions involved in regional planning and development. Thus, for example, Bullman (1997) argues that in the multi-level process of negotiation that is involved in the allocation and use of the Structural Funds, nation-states attempt to negotiate in both directions (upwards with the Commission and downwards with local and regional interests) with the result that dislocation can occur between what may otherwise be considered to be similar or even identical interests. Whilst, as some of the case studies discussed in section three of this chapter demonstrate, this can prove to be a useful learning experience, it can equally be a fraught and a destructive process. One solution is to reformulate the "traditional monopoly of national governments to mediate between domestic and international affairs" (Bullmann, 1997, p.13).

An earlier analysis of difficulties such as these has been provided by Stohr (1989) who pointed to the apparent contradiction which exists between the structure of the organisational system used to manage the regional programmes (a centralised, hierarchical, model constructed in accord with Fordist principles) and the nature of the task which the system attempts to discharge (a decentralised, locally and regionally based task). If anything the current organisational structure could be seen as more centralised than the original 1974 ERDF model.

This introduces a second area of theory, which is concerned with processes of political choice and, especially, the definition and application of subsidiarity. In theory, although not yet in practice, the adoption of the subsidiarity principle should help to mould a new dialogue between the partners involved in the European regional programmes. However, the ability of the present system, in which the Commission and the member states still dominate and attempt to "take care of less developed regions" (Hadjimichalis, 1995 p.97), to restrain the reformulation of the relationships between the regional programme partners, continues to inhibit the final and full adoption of the region as the basic spatial unit for the design and implementation of appropriate restructuring policies.

The final area of theory of particular relevance to the debate on the functioning and future of the regional programmes relates to questions of spatial form and the internal coherence of development programmes. This is an area of considerable research activity in recent years, and the emergence of the European Spatial Development Perspective (ESDP) casts new light on the actual and potential roles and responsibilities of the regional programmes. In particular, the ESDP notes the existence of a

> causal link between the efficient use and leverage effect of Structural Funds and the level of integration of regional development plans into regional or national spatial development strategies (Ministers Responsible for Spatial Planning, 1997, p.40).

The present arrangements (or lack of arrangements) for the preparation and implementation of regional programmes can result in European regional programmes working against, rather than with, other established or emergent regional strategies. In some member states one ministry is responsible for Structural Fund programmes, whilst another ministry has control over other regional planning activities.

These three areas of theory, together with the earlier review of the evolution of the regional programmes supported by the Structural Funds, provide a backdrop against which the performance of the programmes can be evaluated. The issues identified in this section of the chapter also provide criteria against which any proposed reforms can be judged.

Evaluating the Regional Programmes

Introduction

In this section of the chapter an evaluation is presented of the results of two research projects that sought to investigate the operation of the regional programmes supported by the Structural Fund. The first of these projects considered seven programmes located in Scotland, the South West, the West Midlands and Yorkshire. The second project extended the framework of analysis, which had been developed as part of the UK investigation, to four other regions located in Belgium, Germany, Ireland and the Netherlands. These case studies were selected in order to allow for comparisons to be made with the UK experience and for a trans-national analysis to be undertaken.

In selecting the case studies the intention was to:

- provide a range of types of regional programmes within the sample - Objectives 1, 2 and 5b;
- examine the influence of different structures of regional administration and governance ranging from the highly centralised English model to more decentralised structures elsewhere;
- consider different organisational procedures and ways of working;
- identify any particular points of positive innovation that could be applied elsewhere;
- generate evidence that would assist in the overall evaluation of the operation of the regional programmes.

Some of the regional programmes studied operated across two or more NUTS 2 regions. All of the programmes studied operated over the period 1994 to 1996, and some extended beyond 1996.

Three aspects of the development and management of the programmes are considered in this chapter:

- the position of the programmes within the general context of planning, development and management in the regions in which they are located;
- the arrangements for the development of the programmes;
- the implementation and management of the programmes.

Each of these themes is now considered in turn.

Context

The overall context for the programmes can be seen to reflect the general arrangements for regional planning and development in the member states. Wiehler and Stumm (1995) have categorised regional authorities with regard to the degree of autonomy which they exercise; they identify four groups:

- Group 1   Regions with wide-ranging powers e.g. German Lander, Belgian provinces;
- Group 2   Regions with advanced powers e.g. Spanish autonomous communities, Italian regions;
- Group 3   Regions with limited powers e.g. Dutch provinces, Scotland and Wales;
- Group 4   Regions with no powers e.g. Irish counties, English counties.

This categorisation, whilst not incorporating intermediate (mainly non-elected and not directly accountable) arrangements such as the English Government Offices for the Regions (GORs), provides a general overview of the different regimes within which the European regional programmes operate. In practice, these variations in the level of competence which is exercised at regional level are reflected in both the extent to which the European programmes are integrated with other aspects of territorial planning and development, and the amount of autonomy exercised at regional level over the design, negotiation and management of programmes.

The arrangements in Belgium for regional planning and development have been subject to many major changes during the past decade. Restructuring policies have been caught-up in the three classical divides in Belgium society: the socio-economic divide between employer and employees, the territorial divide between Wallonia and Flanders, and the political divide between the various movements. The regionalisation of government in Belgium has caused the Limburg reconversion to be in reality a double reconversion; simultaneously socio-economic and institutional. The first Objective 2 programme (1989 - 1993) was based on the concept of the 1987 Future Contract for Limburg. The primary objective of this Contract was to assist in the restructuring of this declining coal field area with the intention of bringing unemployment in Limburg down to the Flemish average within ten years. In theory, the Future Contract would integrate a variety of funding sources in order to address a series of economic, social and environmental problems.   Innovative financial and administrative mechanisms and institutions were established - the Kempense Steenkoolmijnen (KS,Campine Coalmines) and the Permanente Werkgroep Limburg (PWL, Permanent Study Group) - in order to assist with the implementation of the Future Contract. The Contract itself expressed the

> willingness of all actors to replace the lethargic political-economic landscape in Limburg with innovative decision making bodies (Albrechts et al 1997, p 6).

During the operation of the first programme a number of institutional, financial and other failures occurred. This led to a series of amendments and alterations to, first, the programme itself, and second, the institutional arrangements. Not least of these changes was the need to cope with the effects of the Belgian devolution process.

The previous paragraphs provide a context for the second Objective 2 programme (1994 - 1996). The innovative growth coalition which was dominant in the first programme, and which displaced traditional centres of power (trade unions, employers' organisations, local elected bodies and even the Regional Development Agency), was set aside and replaced by more traditional organisations including the Regional Development Agency and the provincial government. Therefore, the second programme was conceived and prepared in a context of considerable institutional change and uncertainty which

> in Limburg is characterised by the absence of an overall strategy (Albrechts et al 1997, p.20).

In North Rhine-Westphalia, the context for the Objective 2 programme was provided by a much longer established and far more stable set of inter-institutional relationships. According to the constitution, economic restructuring is a

> statutory task of the Land, with the Bund and the EU increasingly playing a supportive role, (Ache and Kunzmann 1997, p.8).

The Objective 2 programme represented one pillar of a complex regional structural policy; the other pillars were, and still are, Bund/Land programmes (support for infrastructure) and Land programmes. Within this general framework a set of 15 regional development strategies were developed for the various regions of North Rhine-Westphalia; these strategies were intended to provide comprehensive concepts for the future development of the regions (Ache and Kunzmann, 1997, p.13)

The Republic of Ireland in its entirety has been designed as an Objective 1 region for the purposes of Structural Fund assistance for the programme periods 1989-93 and 1994-99. Northern Ireland, which forms part of the United Kingdom, also enjoys Objective 1 status. The role of Ireland as both a state and a region for Structural Fund purposes has created a two way tension between a highly centralised administrative system in Dublin (which receives and disburses funds) and the Commission, on the one hand, and sub-national level interests on the other. It can be concluded that by default the thrust of regional policy in Ireland since the reform of the Structural Funds, has been to avail the State of EU Structural and Cohesion funds to assist in bringing the State up to EU levels of economic welfare. This has successfully been achieved, moving Ireland from 63% of average EU per capita GDP in 1986 to 99% in 1995 (Davis, 1997).

The Dutch context for the development of the Friesland Objective 5b programme was provided by long-standing arrangements for strategic planning at the provincial level. Both the Regional Spatial Plans (proposed in 1989 and 1994) and the Socio-Economic Policy Documents (1988 and 1994) made reference to the possibilities of attracting and using European funds in support of development. Existing spatial plans were, therefore, available as a basis for the preparation of a regional strategy in the bid for European funds, even though the agricultural aspects of existing plans were weaker than those elements dealing with economic development, tourism and recreation.

The (European) regional strategy can clearly be seen as a further elaboration on existing regional strategic plans (Zwanikken and Needham, 1997, p.5)

In the UK the situation varied considerably between regions. A series of contrasting strategic inheritances can be identified. The advantages conferred by a single territorial department of central government in Scotland were reflected in the availability of deep-rooted, detailed and comprehensive guidance on various aspects of strategic policy, whilst in some English regions there was, and still is, a more fragmented picture. Although some of the difficulties encountered in the English regions during the preparation of the first round of programmes (prepared in 1988 and operational from 1989 to 1993) were reduced following the establishment of the GORs, it was still the case in 1993/94 that in some regions the context for the preparation of European regional strategies was incomplete or hazy (Roberts and Hart, 1996).

Developing the Programmes

Following the difficulties experienced in the operation of the first Limburg Objective 2 programme, both the Flemish government and a number of other institutional political actors distanced themselves from the preparation of the second programme. In addition, the designated Objective 2 region was smaller than that designated for the first programme - the Objective 2 region for 1994 - 1996 covered 19 municipalities containing 60% of the population as against the province-wide region designated for the first programme. The Regional Development Agency dominated the preparation of the strategy. This body provided a degree of continuity during a period of considerable political turbulence - the Christian Democratic party (CVP) having been ousted by an anti-CVP coalition in 1994. The Flemish and provincial governments negotiated the eventual strategy, which emphasised economic measures and programmes, with the Commission.

In North Rhine-Westphalia the basic structure of the Objective 2 strategy was developed by the Land Ministry of Economics, Technology and Transport (MWMTV), which in terms of concrete projects resorted to the regional development strategies that were developed by the regional assemblies in the fifteen regions of North Rhine-Westphalia. One such body, the Regionalkonferenz Dortmund/Kreis Unna/Hamm, consisted of 72 actors drawn from many different strands of the local arena - local and county authorities, trade unions, chambers of trade and commerce, employers' associations, equal opportunity organisations, welfare bodies, religious groups, colleges and universities, environmental groups and cultural organisations. Through a series of working groups this conference developed and elaborated elements of the regional strategy in collaboration with the Land government.

In Ireland, the regional policy initiatives, structures and spatial plans introduced during the 1960s and early 1970s were largely abandoned during the economically difficult 1980s. In particular, the non-statutory Regional Development Organisations (RDOs), which functioned as representative regional co-ordinating bodies, were abolished by government in 1987. This left a highly centralised state without a developed structure to implement the partnership principle underlying the revised structural fund regulations in the preparation of a strategic plan for submission to Brussels. As a result the Government hastily introduced an ad-hoc structure of seven so called sub-regional areas each served by programme preparation and advisory committees. These were intended to provide sub-national input in the form of regional programmes to the National Plan 1989-1993 then being prepared as an application for the first round of Structural funding. In the event, the plan was submitted to Brussels before the completion of a number of the sub-regional programmes. This confirmed the view that the devolution arrangements were a mere token.

The CSF 1989-1993 provided for the establishment of seven sub-regional Review Committees with somewhat limited powers and restricted membership. The functions of these committees have now been transferred to the recently established (1994) eight Regional Authorities and are exercised through EU Operational Committees. The powers and resources of these new bodies are similarly circumscribed; they likewise remain committees and not partnerships, having merely a co-ordinating and monitoring role at regional level in respect of nationally determined policies. Both the first (1989-93) and second (1994-99) CSF documents reflect the reality of the prevailing Irish administrative structure and national regional policy vacuum.

Whilst the National Plan 1989-93 specified sub-national objectives, the Operational Programme was not disaggregated to a sub-national level in the CSF. Similarly, while the National Plan 1994-99 expresses a commitment to "balanced regional development" the concept is not defined in an operational manner in either the National Plan or CSF.

The regional strategy for Friesland was developed (as noted above) within the context of a series of pre-existing plans and strategies. Collaboration within the Province among local and regional partners led to the preparation of the formal draft regional development strategy and this was then evaluated at national level by a working party representing the relevant ministries. The proceedings of this working party were chaired by the Ministry of Agriculture, Nature Management and Fisheries (LNV). The draft strategy was evaluated and judged in the light of possible inconsistencies with national policies, Commission guidelines and the possibilities for co-financing from national government. Finally, the strategy was agreed with the Commission and a Single Programming Document (SPD) was approved.

In the UK the arrangements for the development of a regional strategy varied in detail from region to region. However, in general, the initial draft was prepared by a group of local and regional partners drawn from local and regional authorities, chambers of commerce, trade unions, employers' organisations, local training and enterprise organisations, universities and other relevant bodies. The draft strategies were then subject to scrutiny and adjustment by GORS (in Scotland this task was undertaken by a special unit within the Scottish Office) and eventually negotiated by central government with the Commission. In earlier rounds of the regional programmes, a criticism was made that central government often imposed changes without sufficient consultation with local and regional partners; this was less evident during the 1993/94 period of strategy preparation.

The Operation of the Programmes

In theory, although not always in practice, the agreed programme (the SPD) flows from the regional development strategy and is therefore the product and the property of a partnership between the Commission, national (and/or regional) government and a range of local and regional actors. Given the varying levels of participation of these partners in the preparation and approval of strategies, together with the considerable differences which exist in terms of the competence of regional government, it is not surprising that the management arrangements for SPDs vary considerably between member states.

As suggested in the preceding section, the second Objective 2 programme for Limburg was a much more modest affair than the 1989 - 1993 programme. The institutional management structure was centred on the Programme Management Unit (PME); this was a secretariat that screened projects and reworked selected proposals in order to ensure their viability. The PME decided if a project should be submitted to the Management Committee (MC). Economic projects were screened by a PME member who was also on the staff of the Regional Development Agency, and social projects were screened by an independent employee of the PME. Economic projects dominated the reconversion programme and, thus, in some cases, "the Regional Development Agency was simultaneously judge and jury" (Albrechts et al, 1997, p.17) since it both submitted and evaluated projects. The majority of social projects were prepared by the Guidance Service for the Mining Area (BLM) which organised vocational training for ex-miners and ethnic minorities. The BLM had the status of a "privileged promoter" together with five other organisations.

The Management Committee for the Limburg programme was composed of representatives from the Flemish government and administration (7), the mining towns (2), the provincial government (2), two scientific experts, five representatives from the PME and two from the European Commission. It was claimed that both the Flemish government and the Commission played a minor role in meetings of the MC. The working of the MC has been described as moving towards the creation of "new power relations and interests in Limburg" (Albrechts et al 1997, p.18).

In North Rhine-Westphalia the Objective 2 programme (which was divided between two areas: the Ruhrgebiet and the Heinsberg area) was basically managed by the Lander with the federal level of government acting as co-ordinator at national level. The Operational Programme (OP) was part of a larger suite of European and domestic programmes in North Rhine-Westphalia . This wider programme, as noted earlier, formed part of a joint task between the Lander and federal government. As such, a federal level monitoring committee had the responsibility for maintaining a general overview of the OP. In operational terms, the detailed management of the programme took place within North Rhine-Westphalia under the supervision of a sub-committee responsible to the Land Ministry of Economics.

Technology and Transport

Application procedures were standardised. Investment projects proposed by the private sector were submitted to a branch of the state bank, the other partners were asked for their comments and the bank then decided on the eligibility of the project for funding. Larger project proposals were submitted to the MWMTV for decision, which again consulted with certain of the partners. Public sector projects were submitted via regional commissioners (in the case of ERDF proposals) or via a decentralised office of the North Rhine-Westphalia Ministry for Labour, Health and Social Matters (MAGS). Regional assemblies were involved in the development of ESF project proposals and, at a first level, in the assessment of projects submitted by third parties.

In Ireland, the Department of Finance has overall responsibility for Structural Fund expenditure supported by a three tier monitoring structure. The Department chairs the national CSF Monitoring Committee, comprising representatives of the Government, Commission and European Investment Bank. Nine national Operational Programme Monitoring Committees with membership drawn from government departments, state agencies and the social partners report to the overall committees. Finally, the EU Operational Committees of the Regional Authorities with membership representative of government departments, social partners, local elected representatives and the community and voluntary sectors make recommendations to the National Monitoring Committees. Evaluation units located within the government departments responsible for the Operational Programmes support the work of the monitoring committees. The CSF mid-term review 1994-96 concluded that the overall CSF process is working well. However, at the sub-regional level, reservations that have been expressed. One regional authority has noted that their participation in the entire process is little more than a form of consultation to placate the EU Commission's suggestion that the principle of subsidiarity is adhered to in the operation of the CSF.

The arrangements for programme management in Friesland involved five organisations. The Provincial Executive was responsible for the management and monitoring of the 5b programme with respect to ERDF and EAGGF elements. The Regional Employment Council was responsible for ESF. A Stafbureau Projecten assisted in the generation and evaluation of projects and the Programma-management was responsible for the daily management of the programme. Overall, strategic supervision was exercised by a Supervisory Commission (Comite van Toezicht or CvT) which included representatives from the partners (European Commission, national ministries, the Province, Regional Employment Council, employers' organisations, trade unions and the municipalities) and which was chaired by the Queen's Commissioner for Friesland. The Provincial Executive made an annual proposal to the CvT on the composition and funding of the programme for the following year and, following the decision of the CvT, this programme was then implemented by the Provincial Executive. During the operation of the programme the Province maintained a direct financial relationship with the European Commission.

In the UK the general pattern of management and operation was that a Programme Monitoring Committee (PMC) superintended the overall programme. The PMC was advised and assisted in the evaluation of proposals by a number of advisory or working groups. The role of the PMC was a strategic one, being primarily concerned with ensuring that implementation was focused on the defined objectives and that these objectives remained relevant. Each programme was provided with administrative support by a Secretariat; in England this function was discharged by the relevant GOR, while in Scotland, Programme Executives undertook this function. The Programme Executives had the particular virtue, in the eyes of some partners, of operating at arms' length from central government. The PMC membership included representatives from local government, training and enterprise organisations, the voluntary sector and the private sector. In general, local authority elected members and the "social partners" were excluded from PMCs.

Conclusions and Future Perspectives

Conclusions from the Research

Trans-national comparisons of the operation of European policy are always difficult to present and justify due to the very different political structures and conventions which exist in member states. However, a number of general observations and conclusions can be drawn from the analysis which has been presented in this chapter. These observations and conclusions relate to:

- the relationship between the European regional programmes and other aspects of regional strategy;
- the extent to which programmes are designed, developed and "owned" by local and regional partners;
- the arrangements for the negotiation and agreement of programmes with the European Commission;
- the formation and operation of partnerships;
- the objectivity and independence of the secretariat and of the management arrangements;
- the effectiveness and accountability of management arrangements, and
- Each of these issues is discussed briefly in the following paragraphs.

As reported in the summary of the case studies, the level of consistency and co-ordination between a particular European programme and other aspects of regional (or sub-regional) strategic policy varied considerably between member states. Whilst in part this is a reflection of the extent of political independence and competence exercised by regional or provincial government, it is also a reflection of the style and the level of evolution of strategic planning. Thus, even though, in Wiehler and Stumm's definition, Dutch provinces do not enjoy the same level of political autonomy as the Flemish government, the deeply-rooted culture of strategic planning in Friesland (and the Netherlands in general) provided a high level of coherence between European and other aspects of strategic development policy. In part the weaknesses in Limburg were due to the considerable political and institutional churning that resulted from the regional devolution process of the late 1980s and 1990s, but this does not fully explain the absence of an embedded technical and institutional structure for strategic planning. Equally, the differences observed between English regions which, in theory, all had the same opportunity to prepare regional strategic plans and to ensure the integration of European programmes with other strategies, cannot be explained solely in terms of political structure and administrative capability. Other factors were at work including the operation of bottom-up regional strategic coalitions and, in some cases, the exercise of strategic initiative by individual civil servants beyond the strict limits of their remit. In Ireland a long tradition of national and regional strategic planning lapsed somewhat during the 1980s. The survival of this planning tradition into the 1990s, and into the European programmes, chiefly took the form of a national strategic plan. In North Rhine-Westphalia the relationship between the European programmes and other "pillars" of restructuring policy reflected the complex and constitutionally-driven nature of the structure and operation of the "one programme - uniform procedures - different funds" method of working. In summary, the extent and quality of strategic coherence would appear to be more a function of philosophy and method of working, than the exercise of political power; and would also appear to reflect the scale and type (productive or destructive) of tension which exists between the various levels of government. The implications of this conclusion are that the inculcation of more advanced methods of strategic planning and management can assist in the effective design and discharge of European and other regional programmes and that, whilst it is not necessary for a region to enjoy full political autonomy (at the level of Wiehler and Stumm's Groups 1 and 2) in order to exercise good strategic planning practice, it is essential that the arrangements for regional governance are guaranteed by constitutional certainty.

The degree of local and regional control over the design and development of regional strategies and programmes is, in part, a reflection of the constitutional arrangements and the extent of autonomy enjoyed at regional level. However, it also reflects the prevailing culture of partnership and strategic working present within a region, either at a formal level or through the operation of informal networking and collaboration. In the case of Friesland, settled constitutional and technical arrangements resulted in the preparation of a strategy and programme that were "owned" by the partnership, by way of contrast, some of the strategies produced by the English partnerships were virtually re-written by central government and programmes were negotiated with the Commission without fully involving local and regional partners. In the case of the Ruhrgebiet many opportunities were provided for local and regional participation, but ultimately real power resided at Land/Bund level. The Limburg case demonstrates the danger of assuming that devolved competence will automatically result in the local ownership of a programme. In the Irish case mature regional organisations were abolished in 1987. Whilst the post

1988 regulations for the Structural Funds required greater local and regional participation, this was not achieved until 1994 and even then in a somewhat token manner.

Because the functions of central government itself are partly devolved in some member states, responsibility for negotiating and subsequently managing regional programmes can take place at a sub-national level. The case of Limburg illustrates the considerable power exercised by regional and provincial government in the process of negotiation and programme management, although this "model partnership" has been somewhat tarnished by subsequent events. The experience of Friesland shows the merits of a considerable degree of regional independence (within a framework of national co-ordination) in the process of programme negotiation and it demonstrates the benefits, in terms of directness and clarity, that result from the province maintaining a direct financial relationship with the Commission during the implementation of the programme. In Ireland despite the belated creation of sub-regional review committees, the CSF was negotiated by central government, which also, in effect, dominated the management arrangements. In the UK case the arrangements for both negotiation and the management of the programme were dominated by central government and the role of local and regional actors was minimal. In the German case, although local and regional actors were involved, the real power would appear to be exercised at Land level; working within established constitutional arrangements.

Partnership arrangements and structures vary considerably between the case study regions. In Germany, Belgium and the Netherlands, although public sector actors dominate many elements of the partnership, and despite the overtly political nature of some of the partnership formation procedures (this was especially true of the Limburg partnership), there appeared to be a genuine desire to represent all interests including the "social partners". In the UK the social partners were excluded from most partnerships, whilst in Ireland, despite the inclusion of the social partners, the partnerships operating as part of the Regional Authorities were in reality more like consultative bodies than true partnerships. The reason for these varying interpretations of partnership is to be found in the Regulations (Commission of the European Communities, 1996) which state that partnership structures and membership will be determined by each member state in-line with national conventions. Overall, the number and institutional representation of partnerships is extending in most of the members states studied, and a growing sense of maturity can be discerned in many of the partnerships.

Objectivity and independence are not always evident in the arrangements for the management of the programmes. This was certainly the case in, for example, Limburg, where the Regional Development Agency provided staff to the programme secretariat who then scored and judged the project proposals put forward by the Agency. However, despite an inevitable degree of duality in terms of the roles discharged by some participants, there is general evidence of an attempt to enhance the openness, accountability and independence of the management arrangements. This process is more likely to be successful if an independent secretariat has responsibility for programme administration and for the evaluation of projects.

The question of effectiveness is more difficult to address in this paper given the absence of a detailed discussion of the research evidence which would allow for a judgement to be made. However, at a general level it is important to distinguish between management/administrative efficiency - many of the English programmes are very efficient at spending European funds - and strategic effectiveness, which is a qualitative measurement of the value added by the European programmes to the overall development of both an individual region and the wider European space.

A View Forward

It is likely that the current arrangements related to the design, approval and implementation of the regional programmes will be subject to a considerable degree of revision prior to the next round of programmes, which is due to commence in 2000. A number of the issues discussed in this paper are likely to figure as important elements in the process of revision. Although the next round of regional programmes is likely to prove to be a transitional period (Colwell, 1997) there are clear signs that the total amount of resource available to the current assisted areas will diminish. In addition, it is clear that the use of the Structural Funds will be expected to serve a wider range of EU policy priorities than previously, and that the level of assistance provided in relation to the total spend within a programme will be recalibrated in order to provide a higher gearing between inputs and outputs. The European Union, like most political organisations, will expect more for its money.

In more specific terms and drawing upon those elements of the research projects that have been presented in this chapter, the lessons from the case studies suggest that:

- there is a mounting body of evidence that the aims, objectives and principles governing the operation of the Structural Funds can best be satisfied by the development and implementation of regional programmes that are fully integrated with other areas of regional development;
- this suggests that the arrangements for regional planning and management should be directed towards the establishment of a single set of planning and strategic development procedures aimed at integrating the European dimension with local, regional and national policy frameworks - this, in essence, suggests the need to establish a clear view of regional development priorities and to prepare an overarching regional development strategy;
- having established a clear strategic vision and plan for a region, which commands the support of the various actors likely to be involved in its implementation, this should form the basis for both European and domestic policy development;
- building on the vision and overarching strategy, a regionally-based partnership should provide the team responsible for developing and agreeing the formal regional programme - the Commission and national governments should not attempt to act as "gatekeepers" or unduly restrict the scope of action of partnerships and the content of the programmes of which they are the "owners";
- given the considerable variations which exist in terms of the constitutional position and the level of competence exercised by regional government, it is likely that a range of models and procedures will emerge - this variety is inevitable and should be encouraged;
- it is also evidence that the quality of management and the transparency of programme operation is enhanced by the establishment of an independent executive or secretariat.

Finally, although not fully discussed in this chapter there are a number of other research findings that are relevant to any discussion of the future development and implementation of the regional programmes supported by the Structural Funds. These include the likelihood that future rounds of programmes will focus attention on specific themes and topics that are of particular European significance, such as competitiveness, environmental enhancement and ecological modernisation, reducing social and economic exclusion, and reducing the level of physical peripherality and what is described as spatial exclusion. In addition, in order to concentrate and ensure the continued effectiveness of a reduced total resource base it is likely that greater spatial concentration and sub-regional targeting will be encouraged. A final priority will be to demonstrate the integrated nature of both urban and (especially) rural programmes of development.

Acknowledgements

The author wishes to thank the following for their contributions to the research which is reported in this chapter: Peter Ache, Louis Albrechts, Guy Beaten, Adrian Colwell, Joe Davis, Gina Finlayson, Trevor Hart, Klaus Kunzmann, Barrie Needham, Greg Lloyd, Erik Swyngedouu, Aiden While and Tim Zwanikken.

References

Ache, P. and Kunzmann, K. (1997), *North Rhine-Westphalia Case Study*, Fakultat Raumplanung, Dortmund: University of Dortmund.

Albrechts, L. Moulaert, F., Roberts, P. and Swyngedouw, E. (1989), *Regional Planning at the Crossroads*, London: Jessica Kingsley.

Albrechts, L., Baeten, G. and Swyngedouw, E. (1997), *Restructuring Processes in Limburg*, Instituut voor Stedebouw en Ruimtelijke Ordening,.Leuven: K.U. Leuven.

Armstrong, H. (1989), 'Community Regional Policy', in J. Lodge (ed) *The European Community and the Challenge of the Future*, London: Pinter.

Armstrong, H. (1994), 'EC Regional Policy', in A.M. El-Agraa (ed) *The Economics of the European Community*, New York and London: Harvester Wheatsheaf.

Bradley, J. (1995), 'Analysing Structural Policies and Growth: Ireland and Portugal', in R.W. Vickerman and H. W. Armstrong (eds) *Convergence and Divergence Among European Regions*, London: Pion.

Bullmann, U. (1997), The Politics of the Third Level, in C. Jeffery (ed) *The Regional Dimension of the European Union*, London: Frank Cass.

Colwell, A. (1997), *Future of the Structural Funds: Interim Position Paper*, Edinburgh: Convention of Scottish Local Authorities.

Commission of the European Communites (1996), *Structural Funds and Cohesion Fund 1994-1999; Regulations and Commentary*, Brussels: CEC.

Davis, J. (1997), *The Republic of Ireland Working Paper*, Dublin: Dublin Institute of Technology.

Hadjimichalis, C. (1995), 'Europe of Regions, Europe of Conflicts', *European Urban and Regional Studies*, 2, pp 95-97.

Hardy, S., Hart, M., Albrechts, L. and Katos, A. (eds) (1995), *An Enlarged Europe: Regions in Competition*, London: Jessica Kingsley.

Michie, R. and Fitzgerald, R. (1997), 'The Evolution of the Structural Funds', in J. Bachtler and I Turok (eds) *The Coherence of EU Regional Policy*, London: Jessica Kingsley.

Ministers Responsible for Spatial Planning (1997), *European Spatial Development Perspective*, Noordwijk: Ministers Responsible for Spatial Planning.

Roberts, P., and Hart, T. 1996), *Regional Strategy and Partnership in European Programmes*, York: Joseph Rowntree Foundation.

Roberts, P. Hart, T. and Thomas, K. (1993), *Europe: A Handbook for Local Authorities*, Manchester: Centre for Local Economic Strategies.

Stohr, W. (1985), 'Regional Policy at the Crossroads: An Overview', in Albrechts, L., Moulaert, E, Roberts, P. and Swyngedouw, E. (eds) *Regional Policy at the Crossroads*, London: Jessica Kingsley.

Vickerman, R. (1992), *The Single European Market*, Hemel Hempstead: Harvester Wheatsheaf.

Weihler, F and Stumm, T. (1995), 'The Powers of Local and Regional Authorities and Their Role in the European Union', *European Planning Studies*, 3,pp.227-250.

Zwanikken, T.H.C. and Needham, D.B. (1997), *Friesland Case Study*, Nijmegen: Vakgroep Planologie, K.U. Nijmegen.